The Tainting of Paradise:
Air Pollution in Belize

We are proof that a growth model can be consistent with a climate resilient and low carbon future of sustainable development.
—Hon. Omar Figueroa
Belize Statement at the UNFCCC COP21
Paris, France, December, 2015

Dedication

This book is dedicated to Belizeans everywhere who genuinely care about their environment and the health of their fellow human beings. It is with hope that the contents of this book will inspire and encourage Belizeans to make wise choices about air pollution and be mindful of the impacts of climate change.

Footprint: A Belizean Environmental Series

The Tainting of Paradise: Air Pollution in Belize is the third book in this series aimed at educating and enlightening Belizeans about the issues affecting their health and environment.

Books in this series:

1—*The Garbage Menace,* Michael F. Somerville, 2015
2—*The Mosquito Book,* Ed Boles, 2016
3—*The Tainting of Paradise: Air Pollution in Belize,* Michael F. Somerville, 2017
4—*GMO or GM No?*
5—*Beyond the Cell Phone: Electromagnetic Radiation and the Invisible Threat*
6—*The Toxic Time-Bomb*
7—*The Energy Debacle*
8—*Intensive Farming and the Demand for the Fowl*
9—*When No More Is Left: Resource Depletion and Exploitation*

The Tainting of Paradise: Air Pollution in Belize

Michael F. Somerville

Footprint: A Belizean Environmental Series 3

Published by *Producciones de la Hamaca*, Caye Caulker, Belize
<producciones-hamaca.com>

ISBN: 978-976-8142-95-5 (print edition)
ISBN: 978-976-8142-96-2 (e-book edition)
The Tainting of Paradise: Air Pollution in Belize is the third in the series:
Footprint: A Belizean Environmental Series ISBN: 978-976-8142-788

Illustrations are by Michael F. Somerville, except Judy Lumb provided the Belize National Protected Areas System map on page 40.

Special thanks to the following sponsors for their contribution towards the development of this book: Belize Audubon Society, Belize Solid Waste Management Authority, Belize Tourism Board, Department of the Environment, Development Finance Corporation, Energy Management Limited, Programme for Belize, ProSolar Engineering, Public Utilities Commission, and Wildlife Conservation Society.

The ideas and concepts in this book are solely those of the author and editors, and do not necessarily reflect the policies and objectives of the sponsors.

This book was printed on-demand by Lightning Source, Inc (LSI). The on-demand printing system is environmentally friendly because books are printed as needed, instead of in large numbers that might end up in someone's basement or a dump site. In addition, LSI is committed to using materials obtained by sustainable forestry practices. LSI is certified by Sustainable Forestry Initiative (SFI® Certificate Number: PwC-SFICOC-345 SFI-00980). The Sustainable Forestry Initiative (SFI) is an independent, internationally recognized non-profit organization responsible for the SFI certification standard, the world's largest single forest certification standard. The SFI program is based on the premise that responsible environmental behavior and sound business decisions can co-exist to the benefit of communities, customers and the environment, today and for future generations <sfiprogram.org>.

Producciones de la Hamaca is dedicated to:
—Celebration and documentation of Earth and all her inhabitants,
  —Restoration and conservation of Earth's natural resources,
    —Creative expression of the sacredness of Earth and Spirit.

Contents

Foreword

I met Michael F. Somerville in 1997 when he was working as Environmental Education Coordinator with the Belize Audubon Society. His interest and passion for conservation was beyond measure. It was no surprise, therefore, that Mike went on to obtain his academic degrees in the environmental field from the University of West Florida. His continued passion for the protection of the environment of Belize has led him to write *The Garbage Menace* which has been well received by Belizeans, particularly because it is well put together, easy to read and understand. Most importantly, it deals with real environmental issues affecting Belize and Belizeans.

This new book, *The Tainting of Paradise: Air Pollution in Belize*, is both instructive and provocative, forcing the reader to think about the day-to-day exposure to air pollution in Belize. As I read it, I could not help but go back during my teenage days when I was involved in burning sugarcane at my father's plantation and how I almost got asphyxiated with smoke when I was trying to extinguish an escaped fire due to a change in wind direction. It also reminded me how, a few years ago, a desperate father requested my advice on how to get a gas station to move from their neighborhood as the exhaust from the fueling vehicles and the gas fumes from the gas station were seriously affecting the health of his asthmatic child. In *The Tainting of Paradise: Air Pollution in Belize*, Mike presents the various types of pollution Belizeans are exposed to, from the cooking with firewood, the black exhaust of old trucks and buses in our city and towns to the exposure of pesticide used for controlling mosquitoes, agriculture pests, to other highly toxic gases and industrial chemicals. Perhaps the best contribution of this book is that it also presents how our life and health is affected by the pollution of our paradise. The final message, however, is that there are still possible solutions to these air pollutions in Belize.

Read on and learn how you are exposed to air pollution, how it can affect your health or that of your family and neighbours, and most importantly, what you can do to ensure we can continue to call Belize a true paradise.

Edilberto Romero
Executive Director
Programme for Belize
May 3, 2016

Preface

My dream of writing an environmental book series based on environmental issues in Belize first became a reality with the publication of *The Garbage Menace*, an educational book that explains the garbage situation in Belize. Like *The Garbage Menace*, *The Tainting of Paradise: Air Pollution in Belize* presents a comprehensive knowledge of the subject area, assembled in one accessible and easy to understand source for Belizeans. *The Tainting of Paradise: Air Pollution in Belize* is divided into four sections: "A Blanket of Precious Air Surrounds Earth," "Where Is Our Air Pollution in Belize Coming From, Anyway?" "How Air Pollution Is Harming Us and the Environment," and "Solutions to Air Pollution." The first section introduces air pollution, and presents a number of air pollution scenarios that could be taking place in Belize. The second section points out the specific activities in our society that are producing air pollution. The third section identifies the various hazardous components of air pollution, and the effects of these on our health environment with consideration to their contribution to global warming and global climate change. The fourth section informs about the many good efforts that are taking place in Belize and other parts of the world to address air pollution and combat climate change, as well as things that we Belizeans can do to help this process. A glossary is provided to further explain unfamiliar words and terms used. A reference list of sources consulted is included for readers who may wish to further investigate some of the topics covered.

In writing this third book, I wish to thank the following individuals: my wife Judith for her continued support of this important undertaking; my editors Judy Lumb and Dorothy Beveridge for their continued enthusiasm and commitment to the series; Dianne Lindo, Tanya Santos, Martin Alegria (Department of the Environment), and Henrik Personn (Caribbean Community Climate Change Center) who took time out of their busy schedules to review portions of the manuscript and provide their professional feedback; and lastly, the many inspiring Belizeans, financial supporters, and other individuals from around the world who are dedicated to protecting the environment and want to see a healthier and better world for all of us to live in.

Michael F. Somerville
April 14, 2016

A Blanket of Precious Air Surrounds Earth

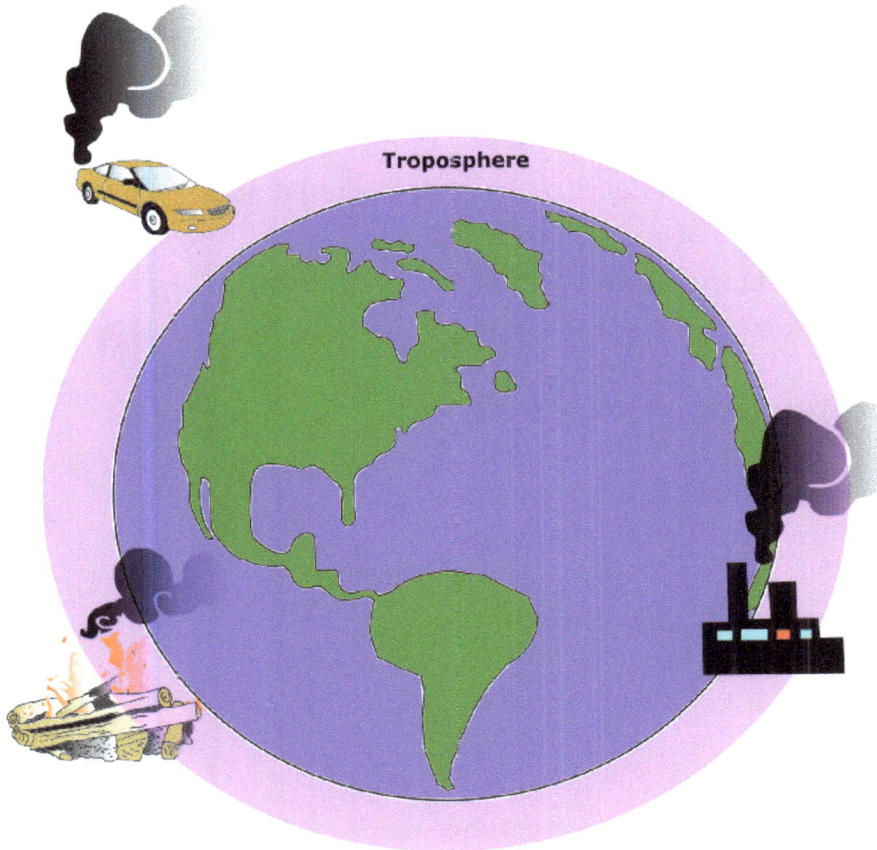

Troposphere

Our life-supporting atmosphere is made up of many kinds of gases in balance with nature.

But what happens when our precious air contains substances, and in quantities, that would not otherwise occur naturally?

Our atmosphere helps to protect Earth and allows life to exist. The Troposphere is the layer of the atmosphere that we live in, and it extends from Earth's surface to about 10 miles above the equator and about 5 miles at the poles.

Putting other gases, fine particles, fumes, smoke, odors, and chemicals into the atmosphere upsets Earth's natural balance and cycles, and can have serious consequences on our health and our environment.

1

All Choked Up...

Have you ever been in a situation where you thought the air you were breathing was dirty or being polluted? How did this affect you and what did you do?

- A passenger bus stirs up a large cloud of dust as it cruises down a street in Belize City.
- The strong odor of paints and thinners reaches people living near a boatyard.
- The smoke from a burning garbage dump drifts through the homes of residents in a village.
- A mother smokes a cigarette in a closed car while her young child plays with a toy in the back seat.
- Inner-city pedestrians are "smoked-out" by the visible smelly exhaust from a poorly maintained passenger "dollar van."
- Birds and insects are attracted to the flame of flares improperly operated at an oil well site.
- Farmers chop and burn the forest upwind of a community.
- A dump truck releases clouds of toxic black smoke as it travels on a highway.
- A family fumigates its home with an insect spray to kill roaches and mosquitoes.
- A lady styles her hair with hair spray each morning before going to work.
- A neighbor burns household trash and tires on his property.
- A cook inhales the smoke from a fire hearth while preparing a meal.
- Lots of smoke and fumes are released through the smoke stacks of a manufacturing company.
- People living near a farm are showered by a cloud of chemicals from a crop duster plane.
- A driver leaves his cargo truck idling while parked at a loading zone.
- Gases are released into the atmosphere from a public garbage disposal site.
- Workers inhale asbestos and lead particles while breaking down an old building.
- On a highway, a driver rolls up the windows of his car to avoid inhaling the toxic black smoke released from a passenger bus up ahead.

Where Is Our Air Pollution in Belize Coming From, Anyway?

The exhaust from motor vehicles is a major source of the poisonous gas carbon monoxide. A variety of persistent organic pollutants (POPs) can attach to fine particles and be transported to regions of Earth far from where they were used or emitted. Smoke from a diesel vehicle contains particulate matter fine enough to enter our blood stream. Air pollution comes from natural events, such as forest fires, wind erosion, dust storms, belching of cattle, natural radioactivity, and volcanic eruptions, as well as the sources described below.

Burning Petroleum

Motor cars are the single biggest air polluters, and dump-trucks and buses are the biggest sources of fine particulates as air pollution.

Exhaust from vehicles that burn fossil fuels accounts for about 13 percent of global greenhouse gas emissions.

Exhaust and fuel evaporation from buses, trucks, cars, motor boats, ships, airplanes, chain saws, lawnmowers, and weed eaters that burn petroleum fuel contain carbon monoxide, nitrogen oxides, volatile organic compounds (VOCs), polycyclic aromatic hydrocarbons (PAHs), sulfur oxides, polychlorinated biphenyls (PCBs), acrylonitrile, fine particles, and persistent free radicals.

Industrial Activities

Manufacturing, Servicing, and Repairing

Power plants, factories, waste incinerators, oil drilling, sawmills, boatyards, and automotive shops put lots of smoke, fumes, persistent free radicals, carbon monoxide, VOCs, PAHs, dioxins, furans, nitrogen oxides, sulfur dioxide, acrylonitrile, styrene, and other chemicals into the air.

Mining and Quarrying

Blasting, crushing and processing of stone and mineral deposits put fine particles, radioactive materials, carbon monoxide, nitrogen oxides, and other chemicals into the air.

Limestone and Dolomite Kilns

Cooking limestone and dolomite puts fine particles, carbon dioxide, carbon monoxide, sulfur dioxide, nitrogen oxides, acidic gases, and heavy metals (lead, arsenic, chromium, mercury, and cadmium) into the air.

Roads, Streets and Parking Lots

Unpaved, dirty, and dusty roads, streets and parking lots put lots of fine particles into the air.

Construction and Demolition

Both construction and demolition of buildings put fine particles, VOCs (formaldehyde, phthalates, etc.), asbestos, lead, and radioactive materials into the air.

Dynamite, Firecrackers, and Fireworks

Exploding dynamite, firecrackers, or fireworks puts fine particles, heavy metals (lead, mercury, arsenic), radioactive materials, dioxins, carbon dioxide, ozone, sulfur dioxide, nitrogen oxides, and hexachlorobenzene into the air.

Noise from fireworks can exceed 140 decibels. Human hearing can be damaged at noise levels of 85 decibels and above. Noise from fireworks may also scare pets and wildlife.

Home, Office, and Commercial Establishments

Fine particles, carbon black, carbon monoxide, methane, nitrogen oxides, ammonia, chloroform, hydrogen sulfide, VOCs (1,3-butadiene, styrene, formaldehyde, butylated hydroxytoluene, perchloroethylene, benzene, toluene, etc.), acrylonitrile, and other chemicals are put into the air by domestic plumbing (including taking hot showers or baths), personal care products (including talcum powder), dry cleaning, paints, varnishes, pesticides, cleaning products, fire-resistant chemicals from furniture and carpeting, spray cans, office machines, gas stoves, barbecue grills, fire hearths, cigarette and tobacco smoke, burning mosquito coils, mold, burning candles, and burning incense.

Smoking

The practice of burning tobacco and inhaling and exhaling the smoke puts many chemicals into the air, and into the lungs of smokers and anyone nearby. Those chemicals include carbon monoxide, tar, nicotine, ammonia, formaldehyde, 1,3-butadiene, and other VOCs, naphthalene and other PAHs, hydrogen cyanide, arsenic, acrolein, DDT, and radioactive polonium-210.

Garbage

Open Garbage Dumps and Landfills

Gases, such as methane, carbon dioxide, nitrous oxide, ammonia, hydrogen sulfide, mercury, VOCs, and other air pollutants are released into the atmosphere from garbage dumps and landfills.

Burning Garbage

When garbage is burned it releases carbon dioxide, carbon monoxide, fine particles, carbon black, persistent free radicals, airborne heavy metals (mercury, lead, chromium, arsenic), PAHs, dioxins, furans, PCBs, VOCs, hexachlorobenzene, and other hazardous substances.

It is estimated that more than 40 percent (over 1 billion tons) of the world's garbage is still burned in open piles.

Agriculture

Animal Farming contributes methane, nitrous oxide, and carbon dioxide to the atmosphere.

Pesticides and Fertilizers released from crop dusters, drones, helicopters, broadcast spreaders, mechanical sprayers, foggers, and manual backpack sprayers put fine particles and chemicals (glyphosate, ammonia, DDT, chlorpyrifos, neonicotinoids, VOCs, etc.) into the air. Tractor plowing may also create dust clouds that contain fine particles and chemicals from pesticide and fertilizer applications.

Slash-and-burn Farming and Burning Agricultural Wastes involves clearing and burning sections of the forests as well as burning the waste from the previous harvest in order to grow crops. This practice puts fine particles, VOCs, PAHs, carbon monoxide, nitrogen oxides, sulfur dioxide, dioxins, and furans into the air.

Burning Sugarcane Fields is the practice of setting sugarcane fields on fire to burn off the leafy material from the sugarcane plant, ward off snakes and other poisonous animals, and make it easier for sugarcane farmers to cut the cane by hand and is usually done before harvesting. This activity puts carbon monoxide, carbon dioxide, nitrogen oxides, formaldehyde and other VOCs, PAHs, dioxins, furans, fine particles, and persistent free radicals into the air. Pesticide residues may also enter the atmosphere if fields were sprayed with herbicides and insecticides.

Roasting Cashew Seeds in pans or drums over an open fire puts toxic vapors (from the burning cashew shell oil), fine particles, carbon dioxide, carbon monoxide, sulfur oxides, nitrogen oxides, and other chemicals into the air.

An annual Cashew Festival is held in the month of May in the village of Crooked Tree.

Deforestation

Cutting down more and more trees for agriculture, industry and urbanization diminishes the forests' ability to absorb and store carbon dioxide from the air. Also, when cut trees rot they give off the carbon dioxide stored in them. The result is an increase in the concentration of carbon dioxide in the atmosphere.

Burning Biomass

Burning materials, such as wood, charcoal, grass, crop waste, or animal dung, produces carbon monoxide, carbon dioxide, nitrogen oxides, VOCs, PAHs, dioxins, furans, fine particles, and persistent free radicals.

Wastewater Management

Sewer systems, sewage ponds, septic tanks, and pit latrines release VOCs, methane, hydrogen sulfide, sulfur oxides, nitrous oxide, carbon dioxide, and chlorine into the atmosphere.

In 2010 Belize had a total of nine sewage treatment ponds, two in Belize City, five in Belmopan, and two in San Pedro Town.

Two-thirds of the country's total households had access to sewer systems or septic tanks while 30 percent used pit latrines.

Eighty-seven percent in urban areas and 45 percent in rural had access to sewer systems or septic tanks, and half of rural households used pit latrines compared to 11 percent in urban.

The District with the highest use of pit latrines was the Toledo District with 57 percent compared to only five percent in the Belize District.

Military Activities

Military practices and trainings put fine particles, greenhouse gases, lead, toxic chemicals, and radioactive materials into the air.

Mosquito Control

Burning mosquito coils, burning grass and other rubbish, and spraying to kill and control mosquitoes put fine particles, PAHs, deltamethrin, malathion, lambda cyhalothrin, propoxur, cyfluthrin, cypermethrin, prallethrin, phenothrin, dichlorvos, tetramethrin, sumithrin, pyrethrins, butylated hydroxytoluene, piperonyl butoxide, formaldehyde, octachlorodipropyl ether, bis(chloromethyl) ether, carbon dioxide, carbon monoxide, nitrogen oxides, persistent free radicals, air borne heavy metals, dioxins, furans, PCBs, VOCs, and other chemicals into the air.

Human-caused Wildfires

Human-caused wildfires may result from discarding cigarettes, burning garbage, leaving campfires unattended, setting fires deliberately to clear land for development, or farming in the slash-and-burn tradition. These fires release carbon monoxide, carbon dioxide, nitrogen oxides, VOCs, PAHs, dioxins, furans, fine particles, persistent free radicals, acrolein, and other chemicals into the air.

How Air Pollution Is Harming Us and the Environment

Globally, about 6.5 million people die prematurely from household and outdoor air pollution each year, with countries mostly in Asia and Africa accounting for the most deaths. This is more than from HIV/AIDS, tuberculosis, and road injuries combined, and makes air pollution the number one environmental cause of diseases worldwide! Air pollution is currently the fourth leading cause of death worldwide after high blood pressure, poor diet, and smoking.

Pollutants and their Effects

Carbon Monoxide exposure reduces the amount of oxygen that reaches our body organs and tissues, and may result in headaches, nausea, vomiting, fatigue, dizziness, weakness, seizure, memory loss, damage to our hearts and brains, coma, and even death.

Carbon Dioxide in the atmosphere causes the greenhouse effect and increasing concentration of carbon dioxide leads to global warming and global climate change.

Fine Particulate Matter exposure may cause mutations of our DNA, heart disease, heart attacks, lung cancer, allergic reactions, bronchitis, and pneumonia. Fine particulate matter may also impair visibility, interfere with photosynthesis in plants by clogging stomata openings, impact ecosystems, and influence climate.

Nitrogen Oxides (nitric oxide, nitrogen dioxide, nitrous oxide) exposure may cause decreased lung function, worsening of lung diseases, and an increased risk of respiratory infections. Nitrogen oxides may also cause acid rain that harms plants and animals, and damages soil, landmarks, and structures. Nitric oxide and nitrous oxide may also deplete the ozone layer, and nitric oxide and nitrogen dioxide may react with other compounds in the presence of sunlight to form ground-level ozone.

Sulfur Oxides (sulfur dioxide and sulfur trioxide) exposure may cause difficulty in breathing, premature death, and increases the risk of lung diseases and respiratory infections. Sulfur oxides may also cause acid rain that harms plants and animals, and damages soil, landmarks, and structures.

Volatile Organic Compounds (VOCs) are given off as gas or vapor from many solids or liquids, and contain carbon, hydrogen, oxygen, fluorine, chlorine, bromine, sulfur, or nitrogen.

Benzene exposure may affect the liver, kidneys, lungs, heart, brain, and reproductive system; and causes anemia, leukemia, and lung cancer.

1,3-Butadiene exposure may result in irritation of mucous membranes, blurred vision, heart disease, birth defects, leukemia, and other cancers.

Butylated Hydroxytoluene (BHT) exposure may interfere with the hormonal system and cause cancer, birth defects, and asthma, as well as behavioral issues in children.

Butylated Hydroxyanisole (BHA) exposure may cause cancers.

Chlorofluorocarbons (CFCs) contribute to depletion of the ozone layer in the stratosphere by the mechanism depicted on the next page. This causes more harmful ultraviolet radiation from the sun to reach Earth. This may lead to skin cancer, cataracts and a suppressed immune system. Ozone depletion can also affect aquatic ecosystems and damage plants. CFC exposure may affect the central nervous system and heart; and causes dizziness, loss of concentration and abnormal breathing (such as choking) in confined spaces due to the vapors displacing air.

Formaldehyde exposure may cause cancer, damage organs, and trigger asthma symptoms.

Methane may lower the normal amount of oxygen in the air leading to suffocation, and increase the average temperature of Earth's surface. It is about 25 times more damaging as a greenhouse gas than carbon dioxide.

Methylene Chloride may damage the eye, cause hepatitis, and cause cancer of the lungs, liver, and pancreas.

Perchloroethylene may affect the brain, increase the risk of Parkinson's disease, and cause cancer.

Styrene exposure may cause memory loss, cancer, headaches, and vertigo; and affect the eyes, skin, central nervous system, gastrointestinal tract, kidneys, and respiratory system.

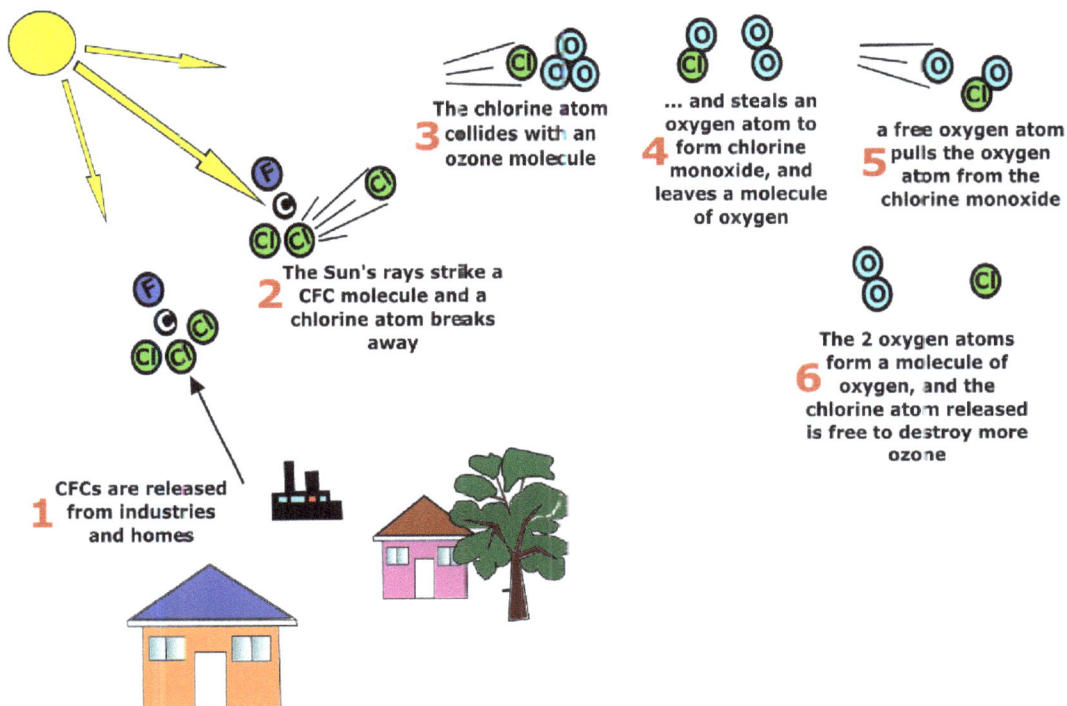

3 The chlorine atom collides with an ozone molecule

4 ... and steals an oxygen atom to form chlorine monoxide, and leaves a molecule of oxygen

5 a free oxygen atom pulls the oxygen atom from the chlorine monoxide

2 The Sun's rays strike a CFC molecule and a chlorine atom breaks away

6 The 2 oxygen atoms form a molecule of oxygen, and the chlorine atom released is free to destroy more ozone

1 CFCs are released from industries and homes

Toluene exposure may cause tiredness, confusion, weakness, nausea, memory loss, loss of appetite, hearing and color vision loss, unconsciousness, and even death.

Xylene inhalation can cause depression of our central nervous systems, headache, dizziness, nausea and vomiting, irritability, slowed reaction time, depression, insomnia, agitation, extreme tiredness, tremors, skin irritation, impaired concentration, and short-term memory problems.

Peroxyacetyl Nitrates (PANs) may affect our lungs, cause eye irritation, mutations, and skin cancer.

Persistent Free Radicals may damage our cells, increase the risk of cancers, heart disease, and stroke; and accelerate the aging process.

Polonium-210 exposure may introduce radioactivity into the body, which could cause lung cancer, DNA mutations, and immune system damage.

Polychlorinated Biphenyls (PCBs) may affect our immune, reproductive, endocrine and nervous systems; damage the liver; and cause breast, uterine, or cervical cancer. Other symptoms include fatigue, headaches, coughs, and unusual skin sores. PCBs also cause deformities and death in wildlife, such as, mammals and birds.

Polycyclic Aromatic Hydrocarbons (PAHs) exposure may cause cancer, mutations, birth defects, lower IQ and asthma. PAHs are also one of the causes of modern-day smog.

Naphthalene exposure may damage or destroy our red blood cells; and cause cancer, confusion, nausea, vomiting, diarrhea, blood in the urine, and jaundice.

Ground-level Ozone can cause inflammation and damage our lungs, make asthma and other lung diseases worse, reduce our immune system's ability to fight off infections in the respiratory system, and aggravate heart disease. Ground-level ozone can also damage vegetation, harm wildlife, and reduce visibility.

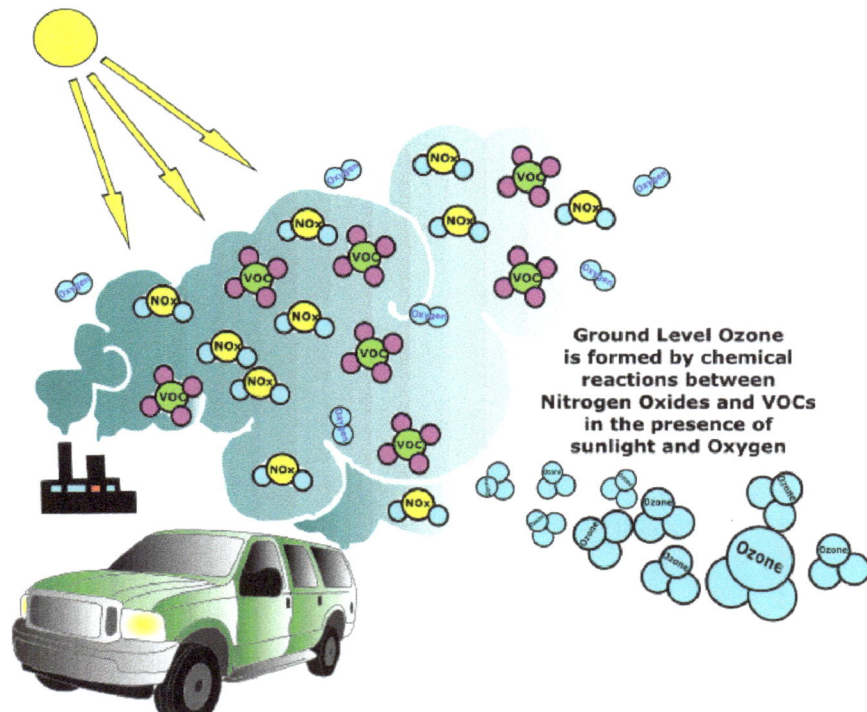

Ground Level Ozone is formed by chemical reactions between Nitrogen Oxides and VOCs in the presence of sunlight and Oxygen

Heavy Metals

Lead can damage our brains, hearts, kidneys, and reproductive and immune systems; lead to learning disabilities in children, high blood pressure and anemia; and poison the environment.

Arsenic may cause cancers and poison the environment.

Mercury can affect our vision, hearing, and speech; cause memory loss, peripheral neuropathy, rapid heartbeat, and high blood pressure; damage our brains, kidneys, and lungs; and poison the environment.

Compact fluorescent light bulbs can release mercury vapors. If a bulb breaks, a good precaution would be to evacuate and air out the immediate area for about 5-10 minutes by opening windows and doors. Cleanup the debris with cardboard or damp paper towel instead of vacuuming or sweeping that could spread the mercury powder or vapor.

Asbestos exposure may cause lung disease and cancer of the protective lining that covers many internal organs, including the lungs.

Carbon Black exposure may cause lung cancer, impact our hearts, and lead to premature death.

Hydrogen sulfide exposure in low concentrations can result in eye irritation, a sore throat and cough, nausea, shortness of breath, and fluid in the lungs (pulmonary edema). Long-term, low-level exposure may result in fatigue, loss of appetite, headaches, irritability, poor memory, and dizziness. Short-term high-level exposure can induce immediate collapse with loss of breathing and a high probability of death. Hydrogen sulfide is also toxic to fish.

Ammonia exposure can affect our skin, eyes, respiratory and digestive systems; displace oxygen in the bloodstream; damage our lungs; and cause death. It is also very toxic to aquatic animals and can react with nitrogen oxides and sulfur oxides in the atmosphere to form secondary particulates.

Pesticides

Acrolein may irritate our skin, eyes and nasal passages, and cause lung cancer.

Chlorpyrifos exposure can affect the central nervous, cardiovascular, and respiratory systems; and cause tremors, loss of coordination, blurred or darkened vision, convulsions, difficulty in breathing, and paralysis.

Cyfluthrin may cause lung problems, convulsions, asthmatic attacks, pneumonia, muscle paralysis, and even death due to respiratory failure. It is highly toxic to fish, invertebrates, bees, and other beneficial insects.

Cypermethrin exposure can cause dizziness, headache, nausea, fatigue, vomiting, skin and eye irritation, seizures, weakened immune system, brain disorders, and possibly cancer. It is toxic to some fish, invertebrates, bees, and other beneficial insects.

Deltamethrin exposure can affect our nervous system, skin and eyes, and can cause headaches and dizziness. Severe exposure may cause nausea, vomiting, abdominal pain, and muscle twitches. This pesticide is also toxic to bees and aquatic organisms, especially fish.

Dichlorodiphenyltrichloroethane (DDT) exposure may cause cancer and damage our thyroid and reproductive systems. DDT is also toxic to certain wildlife and causes eggshell thinning in some birds.

Dichlorvos exposure may cause headache, blurred vision, nausea, vomiting, diarrhea, eye and skin irritation, paralysis, low blood pressure, irregular heartbeat, convulsions, coma, and probably cancer. It is highly toxic to birds, bees, fish, and other aquatic organisms.

Glyphosate (Roundup) is a pesticide that probably causes cancer and may contribute to a higher incidence of ailments, such as gut inflammation and leaky gut syndrome, multiple sclerosis, autism, obesity, depression, infertility, birth defects, Alzheimer's disease, Parkinson's disease, and liver disease. It is also toxic to certain wildlife and aquatic organisms, including amphibians.

Hexachlorobenzene exposure may affect our nervous systems; cause cancers of the liver, kidneys, and thyroid; and cause skin disorders and bone abnormalities, especially in children. This chemical is also toxic to aquatic organisms and bio-magnifies in the food chain.

Lambda Cyhalothrin (ICON) exposure may cause irritation of the skin, throat, nose, and other body parts; and may cause other symptoms, such as dizziness, headache, nausea, lack of appetite, and fatigue. In severe poisonings, seizures, and coma may occur. It is highly toxic to bees, some fish, and aquatic invertebrates.

Malathion exposure probably causes cancer. It can also affect our nervous systems, and cause nausea, dizziness, headache, confusion, convulsions, slowed heart rate, and respiratory paralysis. Its breakdown product, malaoxon, is more toxic. Malathion is toxic to insects, including bees, and to some aquatic organisms, including fish and amphibians.

Neonicotinoids are insecticides that may affect the normal development and function of our nervous system, and damage our brain structures and functions associated with learning and memory. They are also toxic to honey bees, some birds, and other wildlife.

Phenothrin (Sumithrin) exposure may cause skin and eye irritation, nausea, vomiting, throat irritation, headaches, dizziness, and possibly cancer. It is highly toxic to bees, fish, and other aquatic animals.

Piperonyl Butoxide is usually added to insecticides to increase their strength. It can affect our livers, and our nervous, immune, and reproductive systems, and possibly cause cancer. It is toxic to insects, fish, and other aquatic organisms, including amphibians.

Prallethrin may cause irritation of our skin and eyes, headache, dizziness, nausea, vomiting, diarrhea, excessive salivation, fatigue, fluid in the lungs, and seizures. It is also highly toxic to bees and fish.

Pyrethrins are derived from certain types of chrysanthemum flowers and are thus considered to be natural, but exposure may interfere with the normal functioning of the brain and nerves, aggravate asthma, and damage the immune system. These insecticides are also toxic to bees, fish, and aquatic invertebrates.

Pyrethroids are man-made versions of pyrethrins. Studies with lab animals have linked exposure to these chemicals to damage of the thyroid, liver, nervous and immune systems, and disruption of reproductive hormones. They are also toxic to insects (bees and dragonflies), fish, and small invertebrates that form the base of many aquatic and terrestrial food chains.

Propoxur exposure can cause nausea, vomiting, diarrhea, headaches, seizures, respiratory paralysis, and death. It is highly toxic to bees and some bird species, and slightly toxic to fish and other aquatic species.

Tetramethrin can irritate and burn our skin, eyes, and respiratory tract; affect our livers; and possibly cause cancer. It is toxic to fish and bees.

Industrial Chemicals are used in manufacturing chemicals, usually to make polymers. They get into the air as pollutants either from the industrial process or from the breakdown of the polymers.

Hydrogen Cyanide exposure may affect our brains, hearts, blood vessels, and lungs. This chemical can also contaminate food, water, and the environment.

Bis(chloromethyl) Ether exposure may cause lung cancer.

Acrylonitrile exposure may cause lung cancer; skin, respiratory and eye irritation; and may lead to brain and liver damage. It is also harmful to aquatic life.

Flame Retardants exposure may cause infertility and birth defects; liver, kidney, testicular, and breast cancer; and damage to fetal and child brain development.

The Greenhouse Effect

Earth's atmosphere has greenhouse gases, such as carbon dioxide, water vapor, methane, nitrous oxide, and ozone that trap some of the heat re-emitted by Earth after it has been warmed by the sun. This natural greenhouse effect has enabled Earth's temperature to remain stable over time and warm enough for us to live.

Earth's greenhouse effect works the same way as a greenhouse, in which the windows play the same role as the gases in the atmosphere, trapping the heat inside the greenhouse, or like a car in the sun with the windows closed.

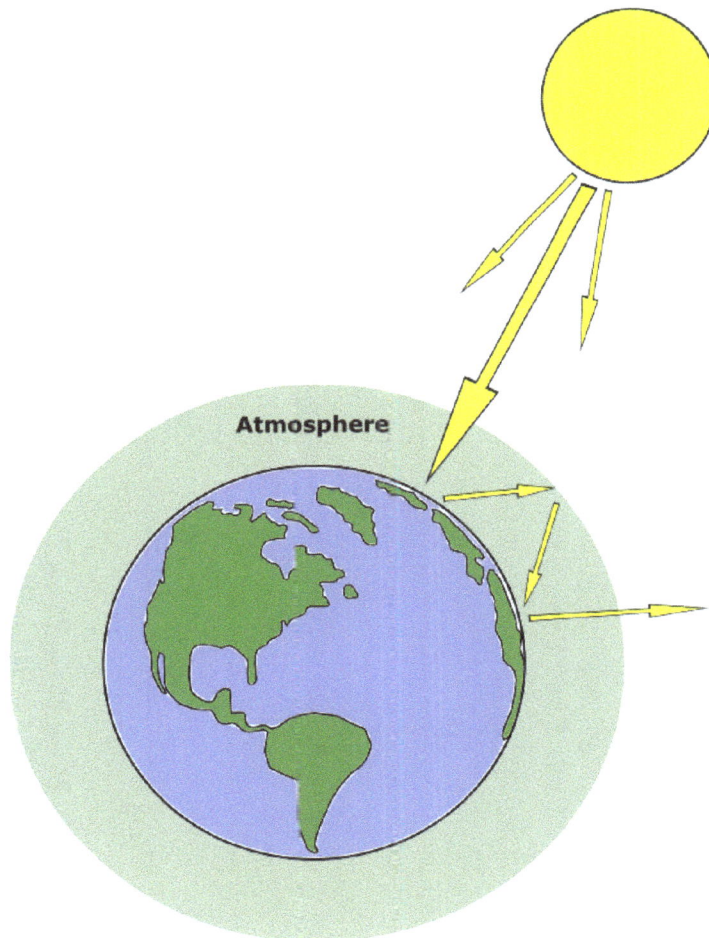

Atmosphere

Global Warming

Human activities are producing greenhouse gases, thus increasing the amount and upsetting the natural balance of these gases in the atmosphere. This higher level of greenhouse gases causes more of the sun's heat to be trapped, making the greenhouse effect stronger and our Earth warmer than normal.

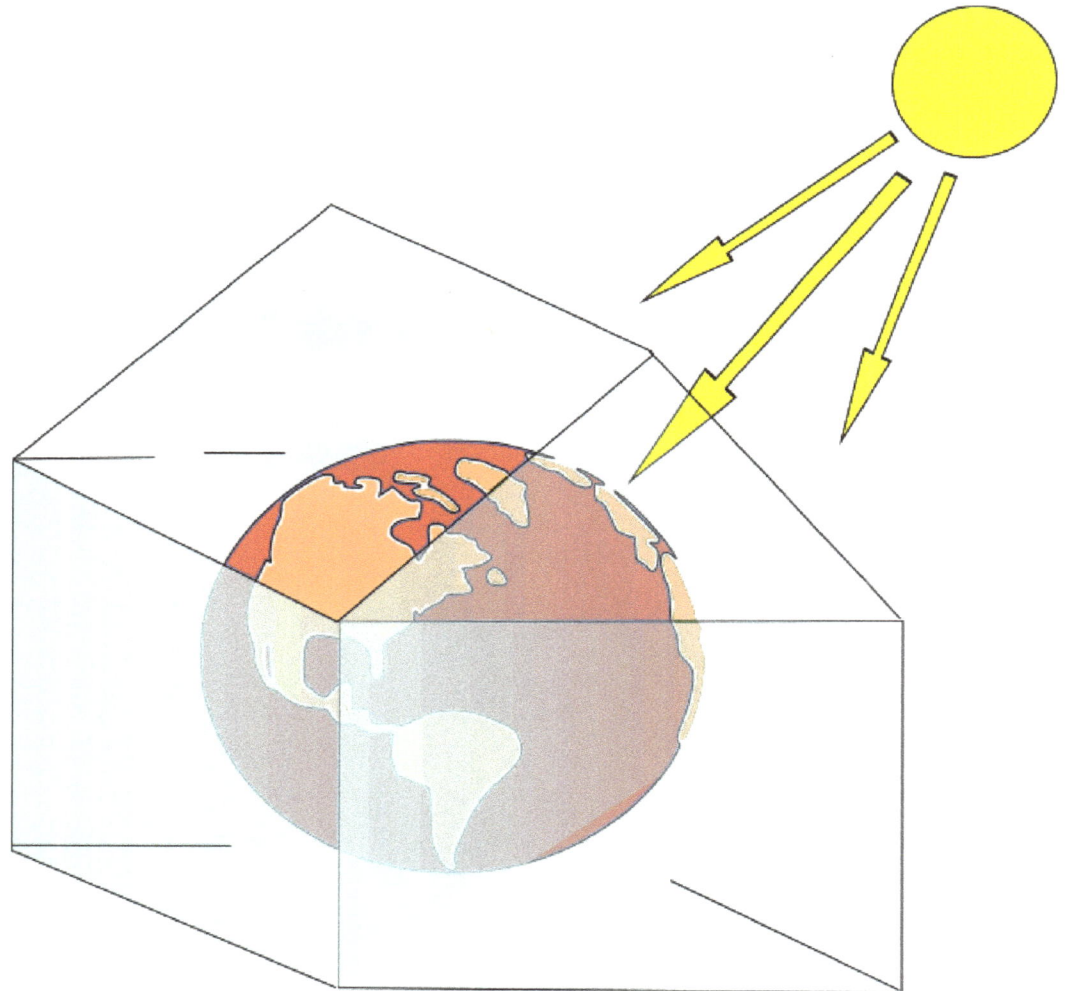

Effects of Global Warming

Global warming is causing the climate all over the world to change. Parts of the world are becoming warmer, cooler, drier, or wetter. Natural weather patterns, such as snow, ice, heat waves, drought, fires, flooding, storms, and El Niño are occuring more often, are more severe, and last longer.

DID YOU KNOW?

In the 1990s, scientists determined that 350 parts per million (ppm) carbon dioxide in the atmosphere was the highest level that was safe to keep global warming to a level that would be conducive to human life on Earth.

The August 2016 carbon dioxide concentration in the atmosphere as this book goes to print is 402 ppm.

Globally, the sea level rise was about 6.7 inches in the last century, but Belize is in an area where sea level rise is greater than the global average.

Although Belize made very little contribution to climate change, we are ranked the eighth most vulnerable to the effects of climate change. More than half of Belize's population and business centers are at sea level.

Climate justice dictates that the developed countries provide support to developing countries in need.

Sea Level Rise occurs from the expansion of sea water as it warms and by the added water from melting land ice (ice caps and glaciers in icy regions of the world and mountain tops).

Ocean Acidification occurs when carbon dioxide from the atmosphere dissolves in seawater forming carbonic acid. Increasing atmospheric carbon dioxide causes oceans to become more acidic, dissolving the stony skeletons supporting corals and reefs. The acidity of ocean water has increased by about 30 percent since the begininng of the Industrial Revolution in the late 1700s.

Coral Bleaching is caused by high water temperatures and increasing ocean acidification. Both cause the symbiotic algae, zooxanthellae, to leave the corals' bodies, causing the corals to turn white (bleached). Bleached corals are weakened, starved of nutrients, and less able to fight diseases. Corals die when bleaching continues for too long. Results from a series of scientific surveys of Australia's Great Barrier Reef reported in April 2016 that 93 percent of the reef showed bleaching effects.

Consequences

Health

There may be increased risk of diseases spread by mosquitoes and other insects. Extreme weather can cause more people to suffer from heat stress, headaches, body rashes, injuries, and loss of life.

Economy and Society

Hurricanes, floodwaters, and wildfires will damage homes and properties. Flooding of vital infrastructure and human settlements could lead to homelessness. Loss of coastal areas and habitats will affect the tourism and fisheries industries.

Agriculture and Food Security

Crops may be damaged by drought and a greater number and diversity of pests. The types of crops that will grow in a particular area will change. Changes in rainfall will also affect the yield of crops that grow in an area, and farmlands and livestock will be ruined by floods. Some countries may not have enough food and many people may suffer from hunger and malnutrition. Loss of food security may, in turn, create havoc in international food markets and could spark famines, food riots, political instability and civil unrest worldwide.

Water

The timing, amounts, and regularity of rain will be impacted. There might be more rain in some countries and less rain in others. Distribution and quality of water will also be impacted. As the human population increases, the demand for freshwater in many countries will increase, causing major conflicts.

Ecosystems and Wildlife

The diversity and value of ecosystems (mangroves, sea grass and coral reefs) could be lost or reduced and many plants and animals might go extinct.

Energy

Increasing human population means increased energy demand for cooling, air conditioning and refrigeration, which will affect demand for electricity, water, and transportation.

Solutions to Air Pollution

As Belize develops, it is possible that air pollution will increase. Since it is very difficult to remove pollution from the air, the best way to solve the problem is to prevent air pollution in the first place. About 19 percent of the world's energy demand was met by renewables in 2012.

Belize Could Introduce More Green Energy

In 2012, Belize's peak energy requirement was about 82 megawatts (MW), and this is expected to increase as the country develops further. In 2010, Belize generated about 60 percent of its electricity from renewable energy sources, such as hydro and biomass (firewood and sugarcane bagasse). Of the remaining 40 percent, 658,319 gallons of fuel and lube oil were used to generate electricity from imported fossil fuel sources (crude oil, natural gas, diesel, heavy fuel oil and imported electricity from Mexico).

Belize can invest in more renewable energy sources to make it more energy independent and resilient, and to minimize the burning of fossil fuels that are polluting our air. Renewable energy is energy that comes from resources like the sun, water, and wind, all of which are plentiful and continuously replenished.

Micro-grid

The Ministry of Energy, Science & Technology and Public Utilities, through a European Union project, is looking at kilowatt-sized facilities to power communities, businesses and public services in Belize. The electricity can be generated and distributed from a centralized generation station powered by solar or other renewable energy sources, such as wind, biomass, and hydro.

Wind Power

Wind power is the fastest growing segment of all renewable energy sources world-wide today. Wind turbines are mechanical devices that convert energy from the wind into electricity.

A single, large wind turbine can stand hundreds of feet high, have blades over a hundred feet long, and produce enough electricity to power hundreds of homes.

A group of wind turbines, called a "wind farm," can be constructed on land that was already impacted by land clearing, and the land between the turbines may be used for agricultural or other purposes. A wind farm may also be located in the sea near the shore or some distance away.

Belize currently does not have a modern wind farm, but the idea of wind power is catching on, and some individuals and establishments in Belize are using small wind turbines to supply some of their electricity needs.

Corporación Centroamericana de Servicios de Navegación Aérea (COCESNA, the organization responsible for air navigation and control for Central America) has a site at Baldy Beacon in Belize that uses two small 7.5 kW wind turbines to generate power for its air traffic navigation beacon and communication system.

Belize Communication and Security Ltd (BCSL) and ProSolar Engineering in Belmopan sell and install small wind turbines.

Solar Power

Although light and heat from the sun may fluctuate during the course of a day or season, solar energy is the most abundant energy resource on Earth. Light and heat from the sun can be converted into electricity, used to cook food, or heat water in a roof-mounted solar water heater.

Solar cells in solar panels convert light energy from the sun into electricity and are used in a number of applications, such as power plants, solar vehicles, water pumps, and in buildings. Panels can be installed on the roof of a home or business and provide all of its electricity needs, or supplement the electricity obtained from the electricity company.

Some companies in Belize, such as ProSolar Engineering and Solar Energy Solutions Belize Ltd., sell and install solar panels in Belize. Some hardware stores sell solar panels.

Most solar power plants use a large array of solar panels. In concentrated solar power plants mirrors or lenses are used to concentrate the beams of light.

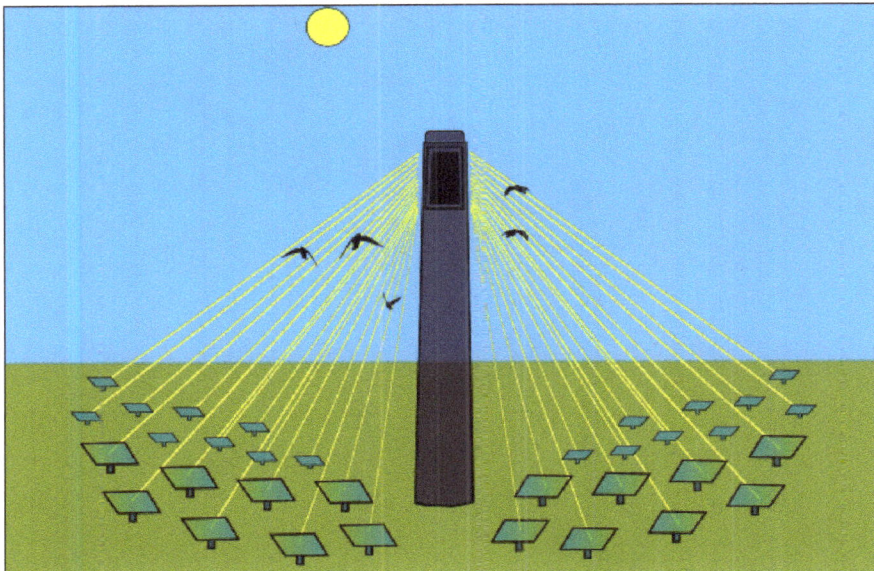

Pico Solar includes small solar panels, rechargeable batteries and LED lights that can be used to provide clean lighting (instead of kerosene lamps and dry cell batteries) and additional essential energy services to off-grid rural areas. Pico-solar can provide power for small electronic devices (cell phones, radios, music players, etc.), be stand-alone kits that can power small appliances, or set up as a small stand-alone power station to run household appliances, businesses or public services.

Solar Stove or Solar Cooker is a smoke-free cooking apparatus that uses a curved shiny surface to collect and concentrate sunlight, and can be used to cook food and heat water. It uses no gas, fuel, or wood, and costs nothing to operate. It helps to reduce air pollution, and lessen deforestation and desertification that can come about as a result of destroying trees to collect firewood. A typical wood-burning stove might produce 400 cigarettes' worth of smoke every hour.

The stove or cooker can come with an energy storage battery or be photovoltaic powered so that it can be used even when the sun is not shining.

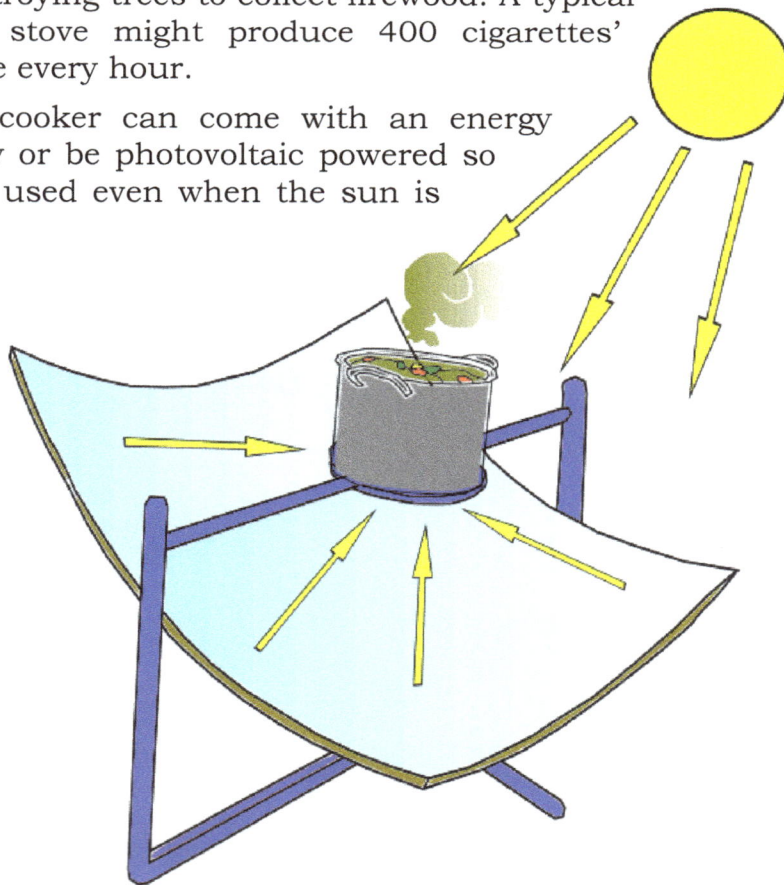

Hydropower

Hydropower is produced in most countries and is currently the most widely used renewable energy source for electricity generation. It uses the gravitational forces of falling or flowing water instead of fossil fuels. Water from a river with adequate flow can be held back by a dam or storage facility and used to drive a water turbine and generator to produce electricity.

A large hydro generally produces over 100 MW of electricity and can power over 100, 000 homes; a medium hydro produces from 10 to 100 MW of electricity and can power about 10,000 to 100, 000 homes; and a small hydro produces from 1 to 10 MW of electricity and can power about 1,000 to 10, 000 homes.

Belize currently has four hydroelectric facilities in operation: Chalillo, Mollejon, and Vaca dams located along the Macal River in the Cayo District; and the Hydro Maya run-of-the-river hydro plant in the Toledo District.

Proposed hydroelectric facilities for Belize include the Chalillo 2 Hydro Project above Mollejon and the Swasey River Hydro Project that includes constructing a hydro facility on the Upper and Lower Swasey River in the Stann Creek District.

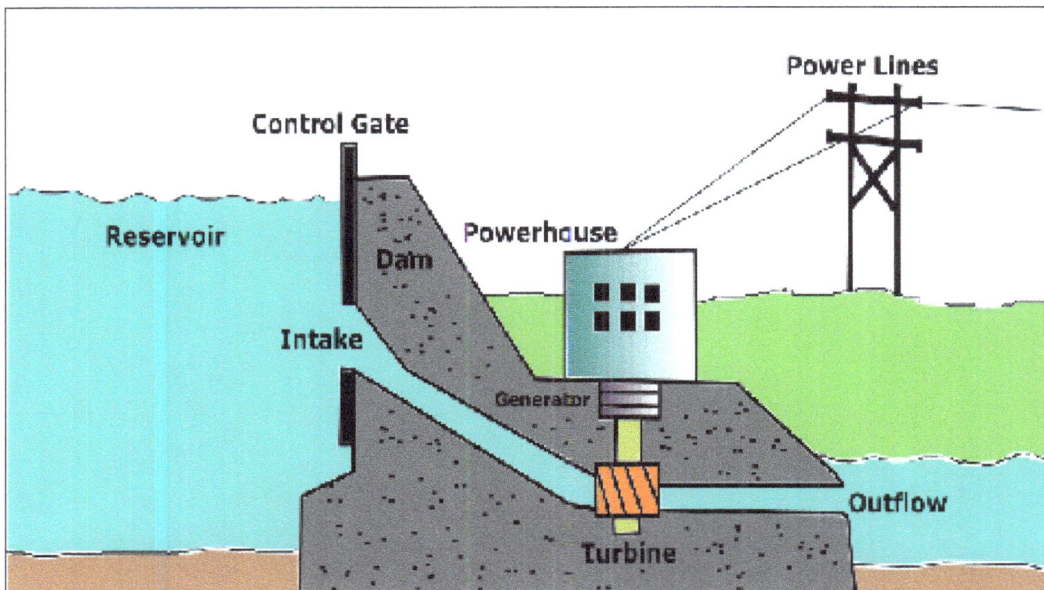

DAMS' NEGATIVE IMPACTS

- Dams may affect the seasonal flooding of rivers and may damage ecosystems.
- Dams may hamper the free movement of aquatic life.
- Dams may block the flow of sediments, leading to downstream erosion and sediment buildup in the reservoir.
- Reservoirs can become breeding grounds for disease vectors.
- Dams may submerge large areas of land forcing relocation of settlements.
- Dam breaks or failures can cause catastrophic damage to settlements downstream.
- Decaying matter from drowned plants and trees in reservoirs can release mercury in the river downstream and generate methane and other greenhouse gases that contribute to global climate change.

27

Mini, Micro and Pico Hydro are run-of-the-river systems that can be installed on rivers, creeks or streams considered too small for larger hydropower plants. They do not generally need a reservoir of water to operate, as they return the water used to the same natural water body. Mini, micro and pico hydros can be located in remote areas and in areas where other technologies would be more difficult to install. They can service small communities, industrial facilities, field stations, lodges, resorts, etc.

Mini-hydro generally produces from 100 Kilowatts (kW) to 1 MW of electricity and can power about 100 to 1000 homes.

Micro-hydro generally produces from 5 to 100 kW of electricity and can power about 5 to 100 homes.

Pico-hydro produces up to 5 kW of electricity and can power up to 5 homes.

ProSolar Engineering in Belize sells and installs micro hydropower systems.

Biomass

Material that comes from living things or things that were once alive, such as dead trees, branches, tree stumps, yard clippings, crop residues, wood chips, saw dust, landfill gas, aquatic plants, and even garbage and sewage waste, is referred to as "biomass".

Fast-growing trees and crops, such as corn, sugarcane, bamboo, wild cane, eucalyptus, and oil palm can be grown specifically to supply biomass.

Biomass can be burned in a power plant to boil water to make high pressure steam that turns a turbine and generator to produce electricity.

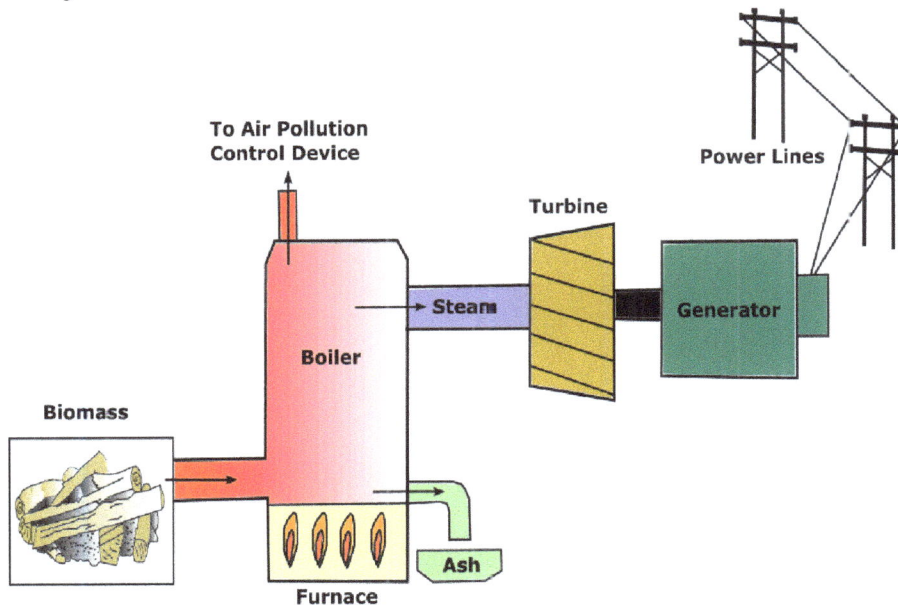

In 2013, the Public Utilities Commission (PUC), Ministry of Energy, Science and Technology and Public Utilities and the Belize Electricity Limited (BEL) issued a call for proposals for alternative sources of energy to move Belize away from its dependence on energy supplied by Mexico, and to satisfy Belize's need for 75 megawatts of energy. Thirty-seven new projects were proposed by 22 companies. Cohune Energy Limited proposed a 15MW cohune energy biomass power plant, which was ranked seventh but put on hold by the PUC.

A BIOMASS POWER PLANT DISADVANTAGES:

- Burning biomass for electricity produces air pollution if the right technology and pollution controls are not in place.
- Biomass burning power plants can use excessive amounts of water for cooling if proper cooling technology is not used.
- Land or habitat degradation can also occur if agriculture or forest residues are not collected properly.
- Land use change or displaced food production can occur if crops are grown on a large scale to supply biomass and sustainable agriculture principles are not applied.

At the BSI/ASR sugar factory in the Orange Walk District, sugarcane residue called "bagasse" is burned at a waste-to-energy power plant to produce electrical energy that helps to supply some of Belize's electricity needs.

The Santander Group's sugar factory in the Cayo District also intends to burn bagasse to supply biomass energy to the national grid. The company milled its first batch of cane in March 2016 and plans to be fully operational by 2017 producing both sugar and molasses mostly for the European market.

Another initiative proposed for Belize includes the GSR Energy Limited Bagasse Co-Generation electricity facility in northern Belize that will use sugar cane produced by local farmers.

Biofuels

Fuels that come from biomass are called "biofuels." Solid biofuels include dried manure, wood, saw dust, charcoal, grass cuttings, etc. Liquid biofuels include ethanol, butanol, biodiesel made from vegetable or animal fat, etc. Gaseous biofuels include methane and Syngas, which is a mixture of hydrogen and carbon monoxide from partially burned biomass. Biofuels are used for transportation, cooking, heating, lighting, and electricity production.

Bioethanol is an alcohol produced by fermentation of sugars from corn, sugarcane, sugar beet, sorghum, cassava, and other sources, including residues and waste from the agricultural, forestry, and waste management sectors. It can be used in automobile engines as a replacement for gasoline. A proposed GSR Energy Limited bio-ethanol distillery in northern Belize will produce ethanol from sugarcane for local use and export.

Cellulosic Ethanol, derived from wood, grasses, or the inedible parts of plants, has added potential to bioethanol because these sources are abundant, and are not in competition with the use of the plant for food, as is corn.

BIOFUELS DISADVANTAGES:

- Growing biofuel crops and plants can displace areas used to grow food and can lead to world hunger and other food insecurities, soil erosion, and deforestation.

- Increased use of biofuels puts increasing pressure on water resources for irrigation, and for boiling and cooling during production.

- Expansion of farming for biofuel production may lead to rainforest and peatlands destruction, loss of biodiversity, pest problems, and increased pesticide use.

- Production and burning of bioethanol and biodiesel may also produce some air pollution.

DID YOU KNOW

On March 11, 2016, United Airlines became the first U.S. airline to start using bio-fuel for its regularly scheduled flights.

BIOGAS VEHICLE DISADVANTAGES:

- Infrastructure for distribution and fueling can be costly.
- CNG and LNG tanks take up more space than gasoline or diesel tanks.
- CNG must be stored in high pressure tanks, and LNG must be stored in specially designed insulated tanks to keep it super-chilled in its liquid form.
- The driving range of a vehicle running on biogas (CNG or LNG) is less than one running on diesel or gasoline due to the lower energy density of biogas.

Biodiesel is produced from oils or fats, such as jatropha seed oil, animal fats, vegetable oil, recycled cooking oil, soy oil, flax oil, sunflower oil, palm oil, coconut oil, peanut oil, hemp, algae, etc. Biodiesel can be used as heating oil or along with conventional diesel fuel in diesel engines.

In 2013, the Department of the Environment facilitated an initiative to recycle used kitchen oils from households and restaurants for blending with conventional diesel fuel. The initiative was discontinued due to lack of cooperation from the public and the costliness of obtaining the used kitchen oils.

Algae can be processed into different types of fuel. The oily part of the algae biomass can be extracted and converted into biodiesel. The green waste left over can be used to produce bioethanol or biobutanol for use in automobile engines as a replacement for gasoline. Biobutanol can be added to diesel fuel to reduce soot emissions.

Jatropha

Biogas is a mixture of gases, mainly methane, carbon dioxide, nitrogen, and hydrogen sulfide, produced by the breakdown of organic matter, such as manure, plant material, or garbage, in the absence of oxygen. Biogas can be used for cooking and heating water, and as a liquefied natural gas (LNG) or compressed natural gas (CNG) in vehicles.

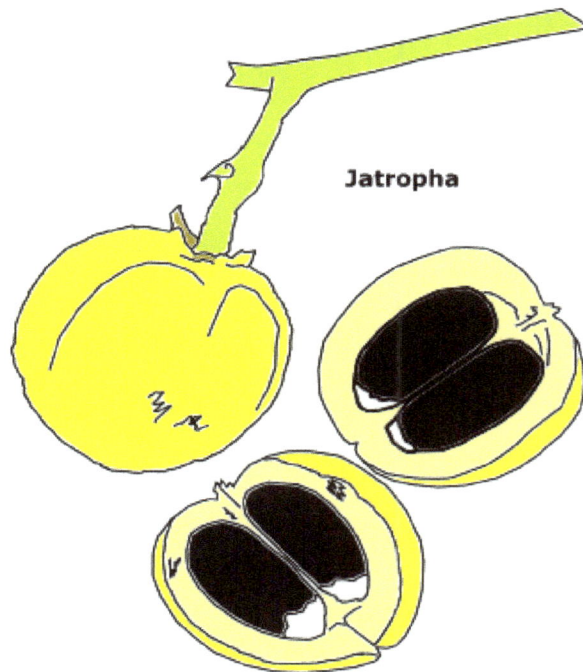

Biogas Digester

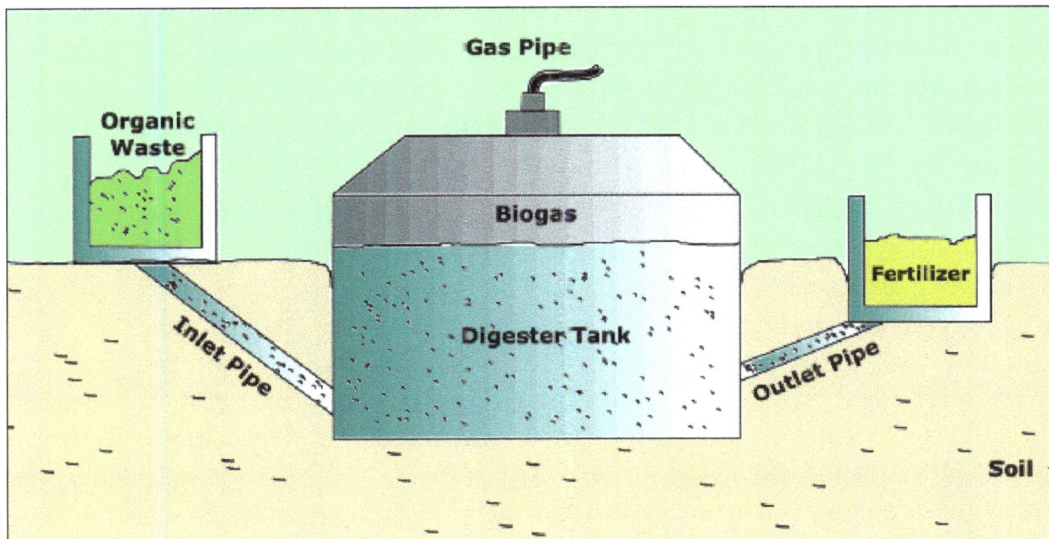

The breakdown of organic waste, including agricultural waste, manure, garbage, plant material, sewage, etc., by anaerobic fermentation in air-tight tanks or drums, called "digesters," produces mostly methane gas that can be used for cooking, heating, running motor vehicles, and generating electricity. The residue after the digestion process can be used as fertilizer.

Belize currently does not have a modern biogas industrial facility, but the Ministry of Energy, Science & Technology and Public Utilities wants to bring this modern technology into Belize and adapt it to local materials in the country.

In 2015, the Caribbean Community Climate Change Center introduced a mobile biogas laboratory at the University of Belize for use within the Caribbean Community (CARICOM) member countries and the private sector. The facility conducts tests on locally supplied feedstock for potential biogas production. The feedstock consists mostly of easy-to-harvest biomass, manure, and organic waste.

In 2016, the CCCCC won the Energy Globe Award for Belize's UB Biogas Laboratory Project. The Energy Globe Award is a prestigious environmental prize recognized worldwide that rewards projects that conserve resources, such as energy or utilize renewable or emission-free sources.

DID YOU KNOW?

If manure is stored at a farm under conditions without oxygen, it may produce high levels of methane and nitrous oxide.

Nitrous oxide is about 300 times more damaging as a greenhouse gas than carbon dioxide and methane is about 25 times more so than carbon dioxide.

Over 9,000 head of cattle were slaughtered in Belize in 2013.

Some livestock farmers and rural households in Belize might have small-scale domestic biogas plants.

BIOGAS DISADVANTAGES:

- Toxic hydrogen sulfide gas might be produced as a byproduct in the process.
- Biogas can be explosive if it escapes from the system and mixes with air.

LANDFILL GAS DISADVANTAGES:

- Gas capture and collection systems can be expensive.
- Gas can be explosive in high concentrations.
- Underground movement can cause organic compounds in the gas to contaminate groundwater.
- Burning landfill gas may produce air pollution.
- Gas can be a source of unpleasant odors.

FLARING DISADVANTAGES:

- Flaring is in the open air, so allows carbon dioxide and other gases, such as hydrogen sulfide and mercury, to escape into the atmosphere.
- Flaring can affect wildlife by attracting birds and insects to the flame.

Landfill Gas

Methane gas produced when anaerobic bacteria decompose organic waste in landfills can be captured and used for cooking and heating, to run motor vehicles, and to generate electricity. Although landfill gas is not considered a renewable energy source, projects that utilize it help to reduce methane emission into the atmosphere.

The Mile 24 Sanitary Landfill is a Clean Development Mechanism (CDM) registered project that captures and eliminates landfill gas through flaring which burns off the flammable gases to avoid buildup and subsurface migration of gas. By capturing and flaring landfill gas, the facility helps to reduce methane gas emissions, and in turn earns certified emission reduction (CER) units (or carbon credits) that can be sold to industrialized countries to meet part of these countries emission reduction commitments. Methane has a stronger greenhouse effect than carbon dioxide, although it is shorter lived. The landfill gas produced at the sanitary landfill will eventually be used to generate electricity once sufficient waste is in place, which will eliminate the need for flaring.

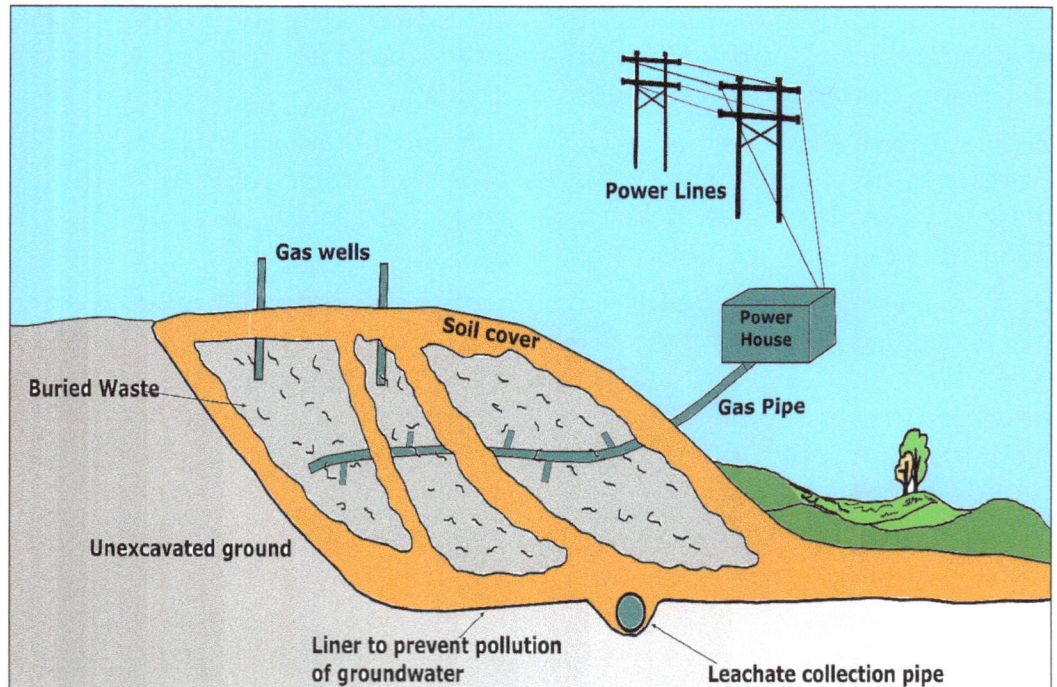

34

Geothermal Energy

The natural heat from beneath Earth's surface can be used for a number of purposes. It depends on the temperature of the geothermal resources (heated water or steam) and how deep they can be accessed. The most active areas for geothermal resources are usually where geysers, hot springs, or volcanoes are concentrated, but geothermal resources that are suitable for heating occur in almost every country in the world.

One way in which geothermal energy can be used is by accessing places where hot water is near to Earth's surface or rises to Earth's surface, such as hot springs and geysers. This water can then be piped directly to homes, offices and businesses.

Another way in which geothermal energy can be utilized is to drill wells deep into the ground in areas where there is greater underground heat or where water circulates at great depth along faults (fractures or cracks in Earth's crust). The very hot water and steam can be piped to the surface and used to turn a turbine and generator to produce electricity.

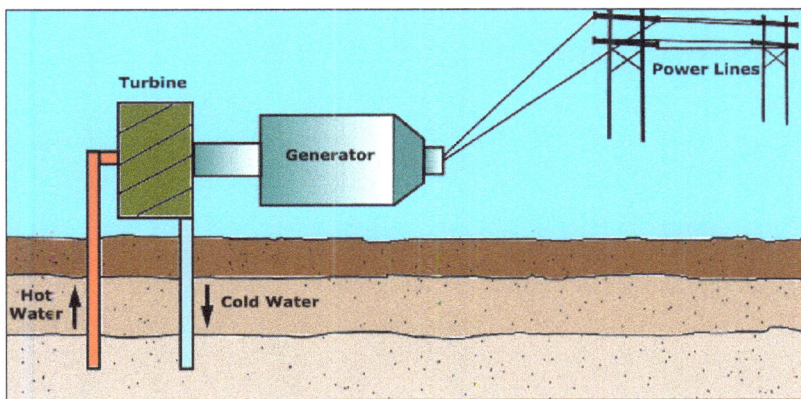

In temperate climates, a geothermal heat pump can be utilized to take advantage of the constant temperature found in Earth about twenty feet down. That temperature is equivalent to the year-round mean air temperature of that region. Water can be pumped from twenty feet down in Earth to air heating systems in the winter and air cooling systems in the summer.

Belize currently does not have a modern geothermal power plant.

Tidal and Wave Power

Energy carried by ocean waves and tides can be used to generate electricity.

Tidal Power is created by gravitational attraction exerted on Earth by the sun and moon that raises and lowers the sea's level. Turbines placed in locations (such as at the mouth of a river) where there are high tidal flows could potentially convert this tidal flow energy into electricity.

Wave Power created by waves moving on the ocean's surface transport energy that could be captured by electricity generating devices. It is believed that these options for generating electricity are not currently viable in Belize from a practical standpoint.

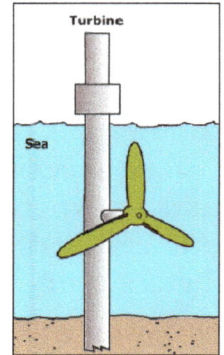

Ocean Thermal Energy Conversion

The temperature difference between cooler (deep) and warmer (shallow or surface) seawaters can be utilized to run a heat engine to produce electricity.

Facilities can be located on the land or near the shore, anchored to the sea bottom of a continental shelf, or as a floating facility off-shore.

The byproducts are useful. The cold water can be used for air conditioning and refrigeration in homes and industries near the facility; the desalinated water (freshwater) can be distilled for drinking; and the deep ocean water can be used to cultivate aquaculture and commercially important species like shrimp and lobsters, and as a source of nutrients and trace elements.

This renewable energy source will probably become more affordable worldwide in the near future, but Belize will probably be the last country in the region to adopt this technology due to the many other options that are becoming available.

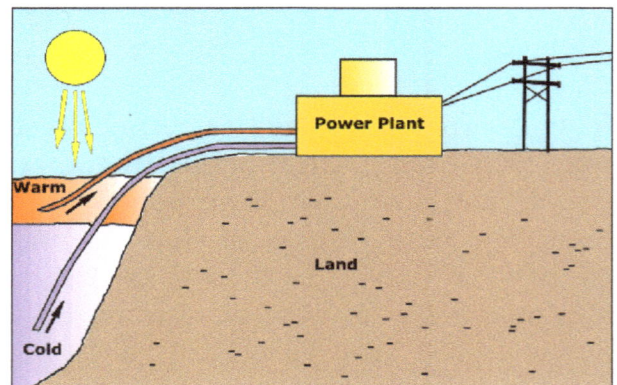

36

Emerging Technology

Hydrogen gas can be used as a fuel, primarily in transportation and industry. Hydrogen gas is very rare in Earth's atmosphere and so is not considered a natural energy source. It can, however, be produced by electrolysis of water using energy from renewable energy sources, such as wind, solar, hydro, geothermal, biomass, etc.

Electrolysis of Water is when an electric current is passed through water to separate water molecules into oxygen and hydrogen.

Hydrogen Vehicles

There are two ways that hydrogen fuel can be used in vehicles. Hydrogen can be burned in an automobile's internal combustion engine instead of petroleum.

Hydrogen is also used in fuel cells instead of internal combustion engines to create electricity to power electric motors in electric cars. Fuel cells are also being developed for use in buses, boats, motorcycles, etc.

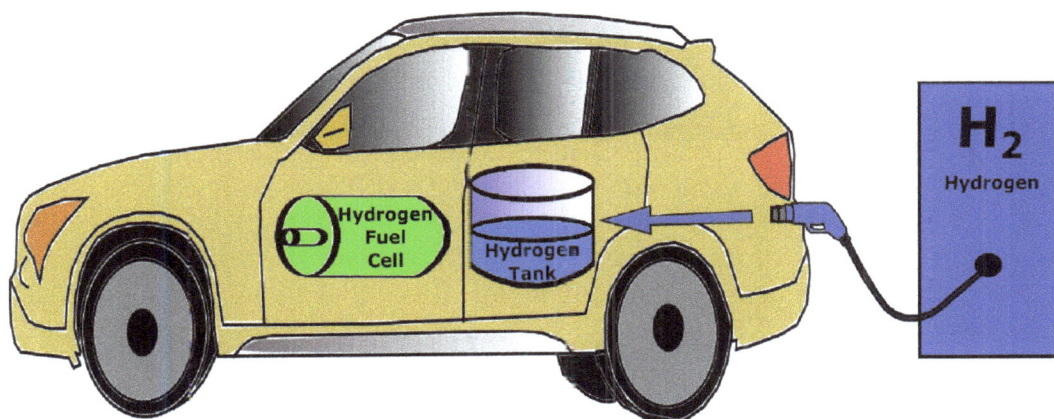

Artificial Photosynthesis mimics the natural process of photosynthesis by capturing sunlight and using its energy to convert water and carbon dioxide into solar fuels. One important solar fuel is hydrogen that can be used for transportation as described above. Progress is also being made in producing other types of carbon-neutral solar fuels, such as methane, methanol, ethanol, etc.

Proposed Technology—Geoengineering

Modifications of natural environmental processes that affect Earth's climate have been proposed to limit the negative impacts of global warming and climate change, but irreversible, unintended consequences could occur so geoengineering is controversial.

Solar Radiation Management (SRM) or Solar Geoengineering

Some of the sun's energy could be reflected back into space before it reaches Earth's surface. Some proposed methods include increasing the

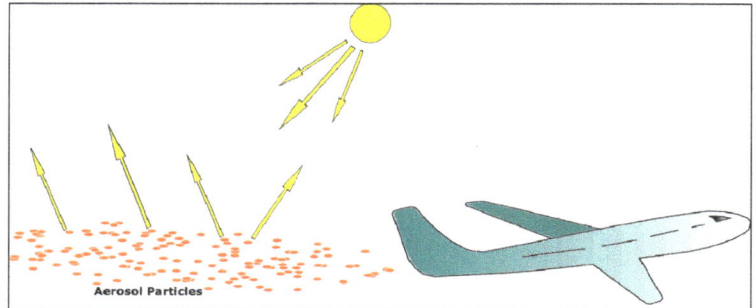

Aerosol Particles

reflectiveness of clouds, roofs, roads, crops, and deserts, placing mirrors in space, and spraying small reflective particles called "aerosols" into Earth's upper atmosphere. A project to pump sulfate aerosols into the atmosphere was proposed in Britain in 2011, but cancelled due to strong opposition from environmental groups.

Carbon Dioxide Removal (CDR) or Carbon Geoengineering

Several methods are proposed to remove carbon dioxide from Earth's atmosphere and oceans. Fertilizing the oceans with iron might stimulate growth of carbon-absorbing planktons. Adding powdered lime would make the oceans more alkaline to increase their ability to store carbon and reduce ocean acidification. Carbon dioxide could be captured from the air and stored. Vast areas, such as deserts, could be turned into forests.

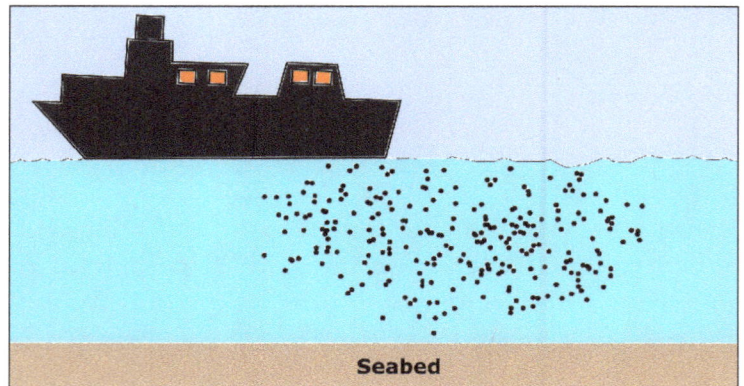

Seabed

Belize Has Good Laws and Initiatives

Belize Laws and Initiatives Address Air Pollution Issues

The Environmental Protection Act empowers the Department of the Environment to prevent and control air pollution in Belize.

The Public Health Act empowers the Minister of Health to make regulations to prevent, control or reduce air pollution.

Belize Climate Change Adaptation Policy mandates the relevant government agencies, including agriculture, coastal zone, education, energy, environment, fisheries, forestry, health, housing, information, tourism, transportation, and water resources, to prepare adaptation policy options for their sectors.

Belize Forest Policy highlights climate change threats and suggests strategies for its mitigation and adaptation through improved forest management practices. It also looks at deforestation and the importance of forests in providing clean air and capturing carbon.

Belize National Sustainable Tourism Master Plan highlights the use of green technology and renewable resources, waste management and recycling, energy efficiency and conservation, improved transportation infrastructure, best practices to minimize environmental impacts to natural resources, and adaptation strategies for climate change.

Belize City Master Plan includes adaptation and mitigation strategies for impacts of climate change and air pollution in Belize City.

National Transportation Master Plan is a sustainable strategy for transportation of freight and people within Belize, and between Belize and its main regional trading partners.

National Land Use Policy highlights climate change adaptation and mitigation strategies for the various sectors.

Belize National Energy Policy addresses reduction of greenhouse gases and air pollution; ways to mitigate the effects of climate change; shifting to renewable energy sources; the need for greater energy efficiency in appliances, buildings, and transportation; and improving energy security and energy access.

DID YOU KNOW ?

In 2010, Belize recorded about 70,535 domestic flights and 7,718 international flights. 39 civil aircrafts were registered, and there were 14,064 total aircraft movements. There were 9,012 departures from the PGIA alone.

In 2016, Sweden opened the world's first electric highway—about 13 miles fitted with overhead power lines that will provide electricity to heavy transport hybrid-electric trucks. When the trucks are off the electric highway they will run on biofuels.

Horizon 2030: National Development Framework for Belize has as one of its strategic priorities the promotion of green energy, energy efficiency, and conservation.

Belize National Protected Areas System provide Belizeans with benefits from ecosystem services, such as food, clean water, clean air, and carbon storage.

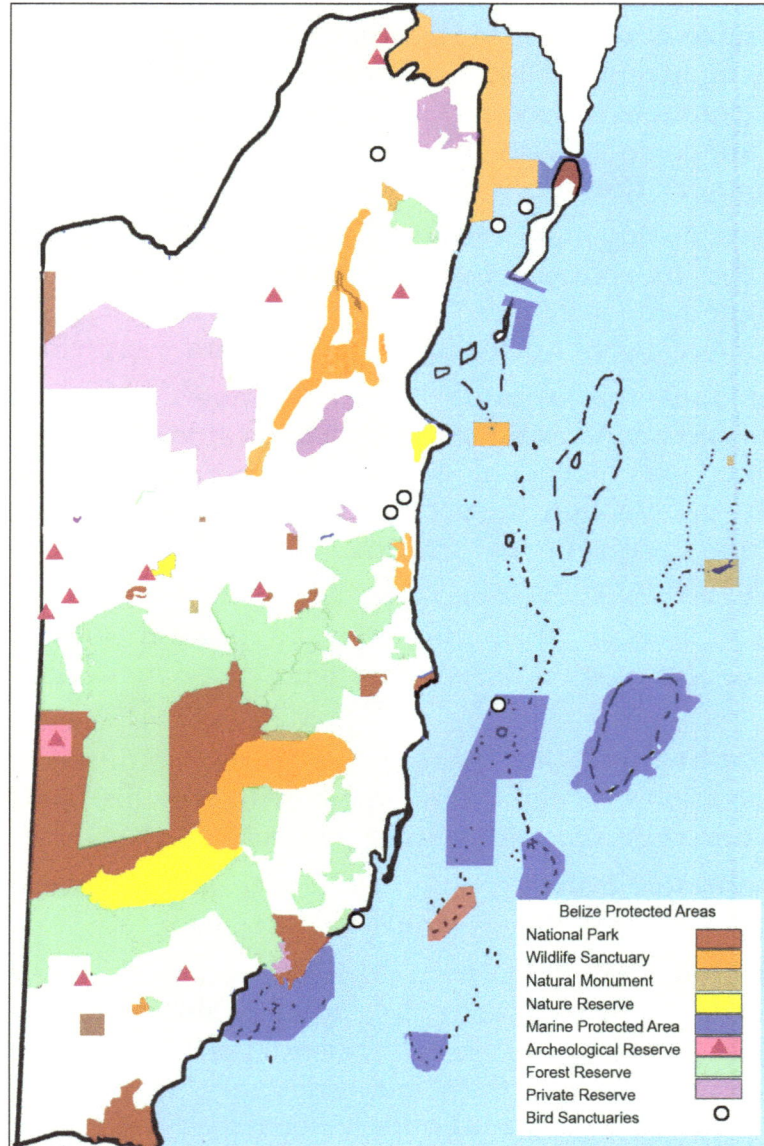

Belize Protected Areas

- National Park
- Wildlife Sanctuary
- Natural Monument
- Nature Reserve
- Marine Protected Area
- Archeological Reserve
- Forest Reserve
- Private Reserve
- Bird Sanctuaries

Belize Participates in International Initiatives

Caribbean Community Climate Change Centre (CCCCC) is headquartered in Belize. The CCCCC works to combat the environmental impacts of climate change and global warming. It provides climate change-related policy advice and guidelines to the CARICOM Member States.

Reducing Emissions from Deforestation and Forest Degradation (REDD) is a United Nations Framework Convention on Climate Change (UNFCCC) mechanism that encourages developing countries like Belize to reduce greenhouse gas emissions by minimizing deforestation. REDD+ goes beyond deforestation and forest degradation, and includes the role of conservation, sustainable management of forests, and enhancement of forest carbon stocks.

Forest Stewardship Council (FSC) is a global forest certification system established for forests and forest products, and addresses deforestation and global warming. Under the FSC's sustainable forestry certification program, forest operations that earn certification can use a seal on their logged wood products so consumers know that the wood they are buying comes from sustainably managed forests. Some organizations in Belize that are members of the FSC include Yalbac Ranch and Cattle Corporation, and the Rio Bravo Conservation and Management Area managed by Programme for Belize.

Small Island Developing States Sustainable Energy Initiative (SIDS DOCK) This initiative supports small island developing states like Belize to transition to a sustainable energy economy through development and deployment of renewable energy resources and promotion of greater energy efficiency.

Ten Island Challenge is a partnership initiative between Carbon War Room and Rocky Mountain Institute to accelerate the transition of Caribbean island economies from fossil fuel use to renewable sources. Belize signed the Ten Island Challenge in 2015, and this move will boost the country's target to be almost 90% renewable in the electricity sector by 2033.

Belize Signed International Agreements

Because many of the challenges contributing to our air pollution problems here in Belize are external, we cannot act alone, but must join with regional and international initiatives to defeat these challenges.

Montreal Protocol on Substances that Deplete the Ozone Layer is designed to protect the ozone layer by phasing out the production and use of the substances (like CFCs) that are responsible for ozone depletion. Belize has a National Ozone Unit.

United Nations Framework Convention on Climate Change (UNFCCC) has as its objective to stabilize greenhouse gas concentrations in the atmosphere at a level that would prevent dangerous anthropogenic (human-caused) interference with the climate system. The Kyoto Protocol that commits countries to reduce greenhouse emissions was adopted in 1997. The Clean Development Mechanism was established under this protocol. The 2015 UNFCCC Conference of the Parties (COP 21) adopted the Paris climate agreement to keep global temperature rise to 2°C above pre-industrial levels. Belize signed on April 22, 2016, and it became effective November 4, 2016.

MARPOL 73/78 includes requirements for the prevention of air pollution by ships.

Basel Convention ensures the environmentally sound management of hazardous waste and control of their movements between nations.

Sustainable Development Goals (SDGs) cover a broad range of sustainable development issues, such as climate change, ecosystems, health, and sustainable and modern energy that Belize has committed to achieving by 2030.

Stockholm Convention on Persistent Organic Pollutants aims to restrict or eliminate the production, use, release and storage of persistent organic pollutants or organic compounds that are resistant to decomposition and can affect human health and the environment.

United Nations Convention to Combat Desertification (UNCCD) combats and mitigates the process of land degradation by which fertile lands become drier and drier—and eventually becoming a desert as a result of climate change, drought, and human activities, such as unsustainable agriculture and deforestation.

Belize Could Improve upon Existing Laws and Make New Laws

Although our existing laws addressing our air pollution issues here in Belize are good ones, sometimes laws need to be improved upon and new ones need to be formulated to adequately deal with current or new situations. These laws would:

- Revise or better implement and enforce existing air pollution regulations to make polluters pay increased fines for damage done to the natural environment.

- Compel manufacturing companies that emit air pollution to install the proper air pollution control devices, equipment or apparatus, such as scrubbers, particulate collectors, bio-filters, etc.

- Ban or phase out residential and commercial open burning of garbage or other combustible materials in rural areas.

- Revise existing air pollution regulations to include emission standards for farm equipment and small gas-powered equipment, such as lawn trimmers and chainsaws.

- Revise existing exhaust emission standards, especially for diesel vehicles and engines, to meet and exceed international compliance.

- Implement a carbon tax on petroleum and other fossil fuels used in transport and electricity generation.

- Revise the Forest Act and the National Lands Act to include specific measures to promote the efficient use of energy and reduce air pollution.

- Better implement and enforce its vehicle speed limits on streets and highways.

- Require that gas stations use vapor recovery nozzles at their gas pumps to prevent fuel vapors from escaping into the atmosphere.

AIR POLLUTION FINES

It is an offence if one burns refuse or other combustible material within any urban area so as to cause a nuisance to any other person. Fines up to $5,000 or a year in prison.

It is an offense to discharge into the atmosphere any contaminant from a gasoline or diesel engine, in excess of the quantity specified by the Minister of the Environment for a motor vehicle operating under normal conditions.

- Institute 'Right to Know" laws for hazardous chemicals that Belizeans may encounter in their everyday lives.

- Make it mandatory for owners of trucks, buses, and water vessels to upgrade or install the necessary particulate and exhaust filters.

- Make it mandatory to place industrial facilities and other sources that emit air pollution at a suitable distance away from any existing or planned community.

- Put the necessary legal and regulatory framework in place to govern the use of renewable energy, including for those systems tied to the electric grid.

- Put mechanisms in place to allow owners of distributed renewable energy systems to sell their excess generation.

- Ensure that renewable energy service providers undergo the necessary skills training, and implement a system to monitor.

- Develop national energy efficiency standards and put in place a certification system for renewable energy and energy efficiency services.

- Institute management standards and building codes for energy efficiency in residential, industrial, commercial and government facilities and buildings.

- Shift the country towards renewable sources of energy for street-lighting, including launching a national program for changeover to more energy efficient lighting in homes and businesses.

- Investigate and implement new sustainable farming methods and technologies that protect human health and the environment.

- Continue to monitor deforestation and minimize or ban unsustainable farming practices (clear-cutting of forests, milpa, cultivation on steep slopes, etc.) that are destructive to wildlife, soil and the atmosphere.

Belize Could Develop Policies and Services

Air Pollution

- Educate and sensitize the general public about air pollution and its effects on human health and the environment.

- Continue to monitor and collect information on its outdoor air quality, and share this information with the general public.

- Setup a national program for rural communities to replace traditional wood-burning stoves with solar stoves for cooking and water heating.

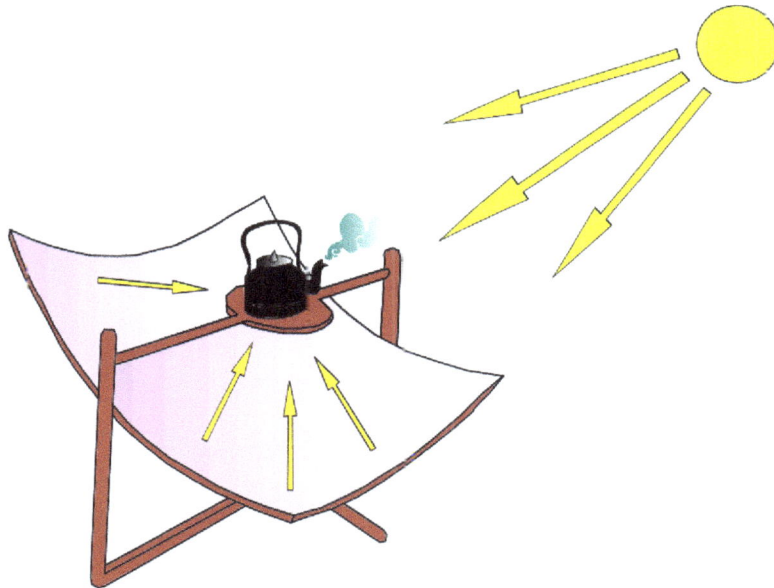

- Institute a program for the collection and proper recycling of fluorescent light bulbs.

- Ban, phase out, or restrict use of dynamite and fireworks, and offer advice on their safe use.

- Consider using the safest chemicals for controlling pests and in mosquito control programs.

Transportation

- Continue to encourage and give incentives (lower vehicle import duties and licensing fees) to individuals who import smaller vehicles that use less energy and emit less air pollution, such as flex fuel, electric and hybrid electric vehicles.

- Consider the environment when replacing the government's fleet of public service vehicles.

- Keep accurate records of vehicle types, kinds of fuel used and year of manufacture to enable better determination of their environmental effects and emissions.

- Improve Belize's mass transport system; encourage or institute carpooling and vanpooling systems to reduce air pollution.

- Restrict heavy vehicle traffic through urban areas, and put measures in place to reduce traffic congestion within cities and towns, especially during peak hours.

- Set up a hotline whereby the general public could be encouraged to report smoking vehicles to the proper authorities.

- Ban or phase out the use of large diesel sport utility vehicles (SUVs), and ban vehicles that are not road-worthy.

- Encourage conversion of vehicles to biogas (LNG or CNG) due to its lower exhaust emissions, overall better safety and less wear on engine parts.

Renewable Energy

- Consider bringing the classroom (that requires lighting and cooling) to students via the Internet, as an alternative to having students and teachers commute to school in vehicles that consume air-polluting fuel.

- Educate the general public about the benefits of renewable energy sources and technologies, and encourage participation of local communities in renewable energy projects.

- Focus attention on a diversity of renewable and resilient energy sources, especially those with fewer negative environmental impacts.

- Increase use of renewable resources for energy production.

- Incentivize the private and public sectors to invest in energy efficient and renewable energy sources for all sectors of the economy.

- Encourage and support small-scale renewable energy systems in communities without access to the national grid.

- Transition to a smart, modernized electricity grid system to handle variable renewable energy generation.

- Explore possibilities with neighboring countries to gain access to cheaper and cleaner electricity from renewable sources, such as geothermal, etc.

- Provide incentives for households to want to adopt greener technology, such as lowering its import duties on energy efficient appliances and on lighting that are energy efficient or utilize a renewable energy source.

- Discourage urban sprawl that requires huge amounts of energy and building material.

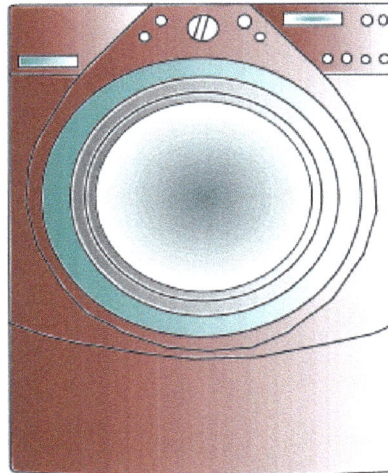

Land Use

- Provide incentives to forest landowners and land users to conserve forests, in the form of payment for forests' environmental services like carbon storage, etc.

- Encourage sugarcane farmers to phase out pre-harvest burning of sugarcane fields and adopt less air-polluting approaches, such as mechanized harvesting and using the leafy portion of the plants as mulch for the next crop or burn it as biofuel.

- Encourage farmers to use better farming practices and technology, and promote farming systems that use mixed intercropping, organic nutrient recycling processes, crop rotations and irrigation facilities.

- Practice good land use management systems, such as agroforestry, through combining trees with crops and/or livestock to maintain forest integrity and environmental health.

- Manage Belize's forests according to the principles of sustainable development, to meet the needs of Belizeans while maintaining the forests in their natural or healthy state.

- Boost the capacity of forest managers to operate at the certification standard of the Forest Stewardship Council.

- Enhance and expand forest management systems to enable effective schemes for controlling deforestation and illegal forest fires.

- Establish forest plantations and initiate or continue to implement national reforestation and afforestation programs that will spare natural forests and increase capture and storage of carbon.

As Belizean Citizens, We Can...

Reduce, Reuse, Recycle, Repair or Refurbish Things

From manufacturing to disposal, every product that we buy might have contributed to greenhouse gas emissions. So, buying less and using less help to lessen the amount of pollution we put into our atmosphere.

- Consider repairing, refurbishing, or recycling an item before throwing it away.

- Buy glass or metal containers for storing food instead of foil or plastic wrap. While plastic containers are also reusable, reusing them too often can cause chemicals in the plastics to leach into our food.

- Buy food items we use often in bulk and those with less or recycled packaging.

- Purchase items that last longer and are easier to recycle, rather than cheap items that break after only a short time of use.

- Minimize buying new clothes.

- Minimize buying items made of leather. If we must buy leather, then buy high quality leather products that will last our lifetime.

- Make a compost pile in our backyard using our fruit and vegetable waste, and use it to fertilize our garden instead of using man-made fertilizers. Manure from farm animals, leaves, wood chips, and sawdust are good, too.

- Use reusable shopping bags and minimize the amount of light weight plastic bags we take home from the grocery stores.

Rethink Transportation

- Consider taking public transportation or join a carpool or vanpool to get to work, to go shopping, or for long journeys.

- For short journeys, consider walking or using the bicycle.

DID YOU KNOW?

Our growing demand for leather products, including wallets, handbags, and shoes requires that more and more massive herds of cattle will have to be farmed, contributing to more greenhouse gas emissions and toxic chemicals required for tanning.

The Green Bag

DID YOU KNOW?

In 2010, there were more than 1 billion cars on the road worldwide, and this number is forecasted to reach 2.5 billion by 2050.

- Consider using the stairs instead of the elevator. It saves energy and it is good exercise, too.

- Buy vehicles and use equipment that are more environmentally friendly and produce less air pollution.

- Keep vehicles and equipment in good working condition.

- Avoid traveling during rush hour or peak hours as traffic jams waste fuel and produce more air pollution.

- Avoid letting cars idle in long lines, such as at a bank's drive-thru, or while waiting to pick up passengers, as this wastes fuel and produces more air pollution.

- Use the right tire pressure to save fuel and produce less air pollution.

- Consider using the air conditioner less as this increases fuel consumption, as opposed to driving with the windows open.

- Avoid warming up our car for an extended period of time as this can waste fuel and produce more air pollution.

- Consider carrying weight at the back of our car instead of on the roof to burn less fuel.

- Driving slower or at a moderate speed burns less fuel and produces less air pollution.

50

Change Our Buying Habits

The further our food has to travel to reach us, the more air pollution is produced. We can find locally grown food at farmers' markets, local fruit and vegetable stands and at some grocery stores.

Go for Organic and Local Produce

- Organically grow our own fruits and vegetables that we eat regularly, such as tomato, sweet pepper, banana, lime, okra, papaya, and watermelon.

- Buy more local and more organic foods, and eat foods that are in season.

- Support local farmers and avoid purchasing fruits and vegetables that are not grown in Belize.

- Avoid eating fast foods and cook our own meals using organic ingredients.

- Consider using coconut oil instead of palm oil for cooking, and minimize purchasing food products or personal care products that contain palm oil, as oil palm plantations are responsible for the degradation and burning of vast areas of forests and peatlands worldwide.

Buy Products That Are Safe For Our Health and the Environment

- Clean and freshen our homes with natural and safe products, such as baking soda, vinegar, and lemon juice.

- Buy environmentally friendly paints, such as water-based paints, that emit less air pollution. Donate leftover paint.

- Minimize buying foam-padded furniture and other items that may contain flame retardant chemicals, and also avoid those furniture treated with non-stick chemicals to repel stains.

- Consider using electric lawn mowers or weed eaters instead of gasoline ones.

DID YOU KNOW?

The Belize Action Community (BAC) organization has started an urban garden project on Zericote Street in Belize City, growing food crops (bell pepper, tomato, okra, onion, etc.) and medicinal plants (moringa, noni, aloe vera, spice, etc.) for local consumption and sale. BAC sees the project as a vehicle for food and financial security for city residents, and plans to train residents and students in urban gardening and plant growing techniques

DID YOU KNOW?

Flame-retardant chemicals are constantly being released from our car seats, baby seats, baby strollers, carpeting, foam mats, couches, mattresses, and other products made with foam.

- Buy personal care products that list ingredients that we can understand—ingredients with long, difficult-to-pronounce names are usually harmful.

- Avoid any kind of air fresheners and deodorizers containing fragrances that are not natural.

- If we must use mosquito coils indoors, keep them in well-ventilated areas and avoid sleeping near them while they are burning.

- Consider using a mosquito net instead of chemicals.

- If we have to burn candles, avoid those that are scented or made with paraffin, and use those made with pure beeswax that produce less soot and air pollution.

- Use pesticides made from natural sources to treat insect pests on our farms, garden plots, and in and around our homes.

- Add live fish to open containers of water or areas of standing water in our yards to eliminate the larvae of disease-carrying mosquitoes.

Conserve Energy

We may not have a lot of control over how our energy is produced, but we can control the way in which we use it. By using less energy, less of it will need to be produced and less pollution will be released into our atmosphere.

Save Electricity

- Consider using a roof-mounted solar water heater in our home instead of a gas water heater or electric showerhead.

- GreenSun, Ltd. is a company in Placencia, Stann Creek District, that sells and installs solar water heaters.

Warm Water Storage Tank

Hot Water to House

Cold Water In

Solar Collector

Roof

- Consider painting our house a light color to keep it cooler.

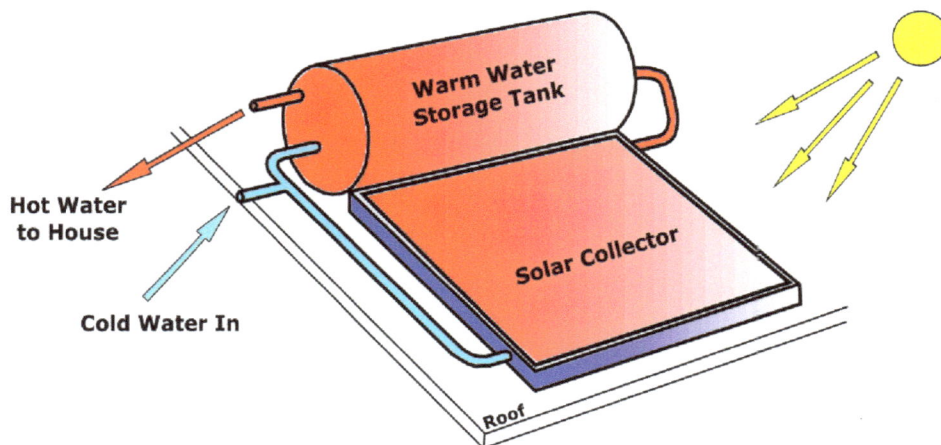

- Switch off or unplug our electronic devices when not being used as these continue to use electricity in stand-by mode.

- Buy energy-saving items (light bulbs, household appliances, office equipment).

- Turn off the lights and appliances when they are not being used.

- Unplug our phone charger when not charging a phone.

- Buy rechargeable batteries for devices we use frequently.

- Set up our work or study space near a window to maximize use of natural light.

Conserve Water

A lot of energy is used to pump, treat, clean, and heat water for our every-day use, so using water wisely saves energy and reduces greenhouse gas emissions.

- Avoid running the water continuously when washing dishes, and lessen our use of automatic dishwashers that can use large amounts of water.

- Turn the water off while brushing our teeth or shaving, and take shorter showers.

- Fix a leaky faucet that can waste a lot of water over time.

- Wash only full loads of laundry, and use cold water instead of warm or hot.

- Use water-efficient appliances and plumbing fixtures.

- Use more water from our vat or rain barrel to cook, wash dishes, and bathe with instead of using from the pipe, and use the waste water after washing dishes to water our yard plants.

- Use a water bucket to wash our car instead of using the hose that can waste lots of water.

- Avoid disposing of dangerous chemicals or substances down the drain, in the toilet, or on the ground, as these could poison surface water and groundwater.

Help the Ozone Layer

- Replace and properly dispose of our old refrigerator (if it was made before 1995) and our old air conditioner (made before 1994) as they probably use chlorofluorocarbons (CFCs).

- Minimize use of air conditioning in our homes.

- Ensure that our vehicle's air conditioning system is working properly and not broken as this could leak CFCs.

- Minimize using foam packaging materials as many of these contain CFCs and HCFCs.

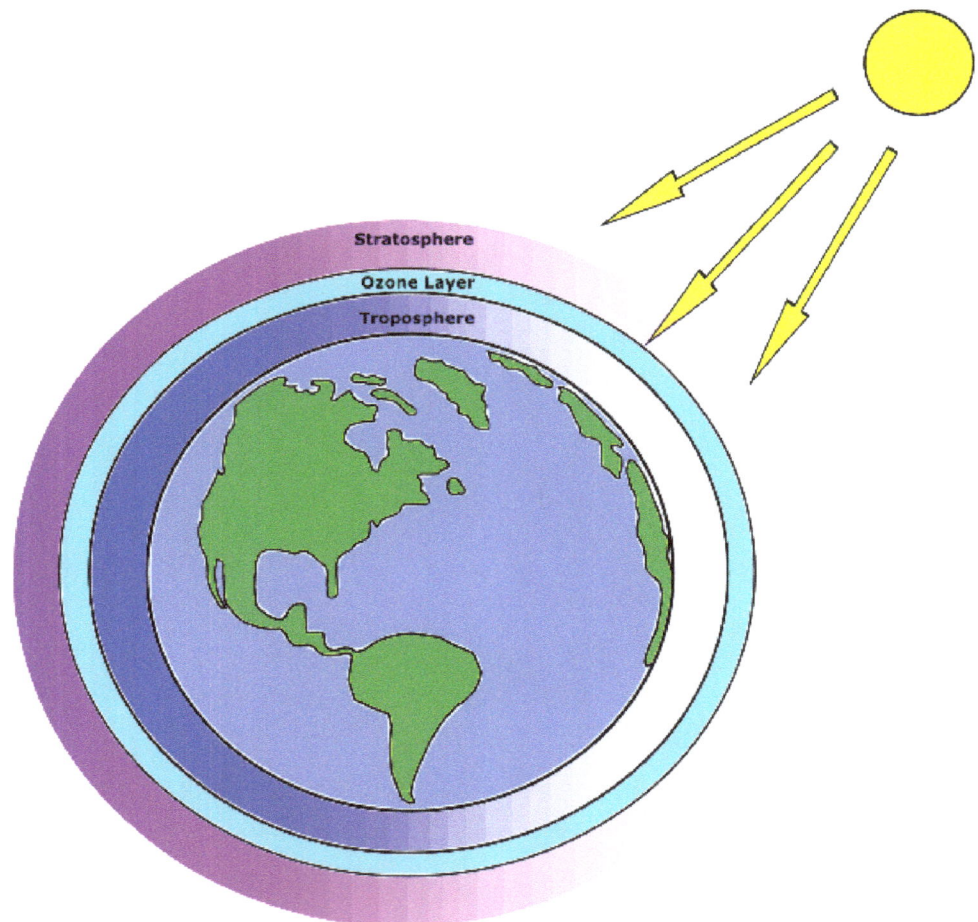

Stratosphere

Ozone Layer

Troposphere

Act Responsibly

- Educate ourselves about air pollution and be wary of air polluting sources in our environment.

- Join an organization or group working to reduce air pollution.

- Help to spread the word about air pollution through donating a copy of this book to a friend or family member.

- Avoid smoking, and urge our family and friends to do so as well. If we must smoke, do so in a well-ventilated area and away from others.

- Store hazardous chemicals properly and in a safe location at our home and workplace to avoid contaminating ourselves, others, and the environment.

- Minimize pest problems by keeping our home and surroundings clean and well maintained.

- Conserve and avoid wasteful and extravagant use of resources.

- Report reckless and suspicious air pollution activities to the authorities.

- Consider building your next home with wood instead of cement and steel. Wood is a renewable resource and its products store carbon. It also requires less energy and fossil fuels to extract, produce and transport, and it emits less greenhouse gases and release less air and water pollutants.

- Plant trees in our yard and neighborhood to provide shade, reduce the demand for air conditioning and cooling, and reduce air pollution.

- Keep houseplants, such as aloe vera, chrysanthemum, bamboo palm, rosemary, lavender, mint, jasmine, geranium, ficus, flamingo flower, English ivy, and peace lily in our home and workplace to absorb pollutants and help purify the air.

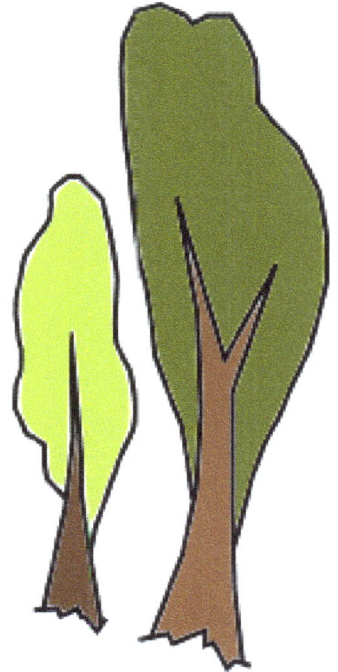

Afterword

We are all exposed at some point or the other to the fallout from a burning garbage dump; the smelly fumes from boatyards, automotive shops and poorly-maintained vehicles; smoke from factories and wood stoves; and noxious chemicals used in our never-ending battle with insects, weeds, and pest animals. While air pollution in Belize might be perceived as moderate or even low compared to other countries, such as Mexico, United States, China, and India, it should not be ignored. The effects of air pollution are legion and may include heart disease, respiratory disease, reproductive disorders, sterility, brain damage, cancer, and loss of ecosystems and wildlife. As our country continues to develop it is likely that its current level of air pollution will increase if effective measures are not put in place. Today, carbon dioxide (the greenhouse gas that plays the biggest role in global warming) is believed to be at the highest level in Earth's atmosphere that it has ever been in the history of mankind, and further increases may result in catastrophic consequences to our world, including sea level rise, food insecurity, more extreme storms and an increased risk of diseases. But even with the eventual decline of fossil fuels and the seeming increase in the use of renewable sources of energy by world governments, the quality and cleanliness of our air remain a responsibility for each of us.

Glossary

1,3-butadiene: A non-methane volatile organic compound used mainly in the production of synthetic rubber, such as that used in automobile tires.

Acid Rain: An unusually acidic rain or any other form of liquid or solid water particles (snow, sleet or hail) that falls from the atmosphere and reach the surface of Earth.

Acetone: A colorless liquid that evaporates, burns easily, and is mainly used to dissolve or loosen another substance.

Acrolein: A toxic substance with a disagreeable odor mainly used as a biocide.

Acrylonitrile: A highly flammable and toxic colorless liquid used in the manufacture of some plastics.

Aerosol: A mixture of fine solid or liquid particles in a gas.

Afforestation: The establishment of a forest by planting trees on open land that has been without forest for a long time.

Algae: A group of single-celled or multi-celled simple plant-like organisms, including seaweeds, diatoms, and spirogyra.

Ammonia: A colorless gas with a strong, sharp smell usually given off from agricultural processes, fertilizers, cleaning products, and explosives.

Anaerobic: Without air or oxygen.

Antimony: A silvery-white, brittle substance with both metallic and non-metallic properties found naturally in Earth's crust.

Arsenic: A naturally-occurring heavy metal used in pesticides, paints, and for preserving wood.

Asbestos: A mineral fiber found in rocks and soil, used widely in the automotive and construction industries.

Benzene: A volatile organic compound found in cigarette smoke, car exhaust and petroleum, and used to make products including plastics, lubricants, dyes, detergents, and drugs.

Bio-magnification: The tendency of pollutants to increase in concentration in organisms at successively higher levels in a food chain.

Butanol: A flammable alcohol used primarily as a solvent and in the production of industrial and consumer products.

Butylated Hydroxyanisole (BHA): A chemical used as a preservative in rubber and petroleum products.

Butylated Hydroxytoluene (BHT): A chemical used to preserve fuel, oils, and other materials.

Cadmium: A naturally-occurring heavy metal used mainly in the production of batteries, pigments, metal coatings, and plastics.

Cancer: A disease caused by abnormal and uncontrolled cell division.

Carbamate: A chemical substance used in the production of pesticides to kill insects.

Carbon Black: A black material produced by the incomplete burning of petroleum, and used to reinforce rubber and in the manufacture of pigments, ink, paints, and crayons.

Carbon Credit: A tradable certificate or permit that shows that a government or company has paid to have a certain amount of carbon dioxide or the mass of another greenhouse gas with a carbon dioxide equivalent removed from the environment.

Carbon Dioxide: A colorless, odorless gas that is present in the atmosphere and formed during combustion of fuel containing carbon.

Carbon Monoxide: A colorless, odorless, and tasteless gas that results from the burning of natural gas and other materials containing carbon, such as, gasoline, kerosene, and wood.

Carbon Neutral (Net Zero Carbon Emission): Removing as much carbon dioxide from the atmosphere as is released into it.

Carbon Sequestration: The removal and storage of carbon from the atmosphere in carbon sinks, such as oceans, forests or soils through physical, chemical or biological processes.

Carbon Tax: A tax charged to carbon emitters that produce carbon dioxide through burning petroleum or natural gas that is aimed at reducing the production of greenhouse gases in the atmosphere.

CARICOM: Caribbean Community, an organization of Caribbean nations.

Carpool: An agreement whereby many individuals travel together in one vehicle, usually a car, and share the costs.

Central Nervous System: The part of the body comprising the brain and spinal cord that regulates responses to internal and external stimuli.

Chikungunya: A virus carried by infected *Aedes aegypti* or *Aedes albopictus* mosquitoes that causes an illness, the symptoms of which are high fever, rash, headache, nausea, fatigue, joint and muscle pain and swelling.

Chlorofluorocarbons: Volatile organic compounds containing chlorine, fluorine, and carbon, used in the production of aerosol sprays, foam insulation, cleaning products, and as refrigerants in refrigerators, freezers, and air-conditioning systems.

Chlorine: A chemical gas or liquid with a suffocating odor, primarily used for bleaching, disinfecting, and purification.

Chlorpyrifos: A chemical used to kill insects.

Chloroform: A colorless, toxic, sweet-smelling liquid, used mainly as a solvent, and once used widely as a general anesthetic to bring about unconsciousness.

Chromium: A naturally-occurring heavy metal used for chrome plating, dyes and paints, tanning leather, and preserving wood.

Civil Unrest: Riots, sabotage, and other forms of crime caused by a group of people as a form of protest or displeasure.

Co-generation: An efficient means by which a power station produces electricity and useful heat at the same time.

Combustible: The ability of a substance to catch fire and burn.

Compost: Organic matter that has been broken down by organisms and used as a natural fertilizer for growing crops and other plants.

Contaminant: Something that makes a place or a substance unusable.

Continental Shelf: The edge of a continent that is situated under water.

Crop Rotation: Growing different types of crops in the same area at different times to control insects and improve soil fertility.

Cyfluthrin: A pyrethroid chemical used to kill insects.

Cypermethrin: A pyrethroid chemical used to kill insects.

DDT: A chemical substance once widely used to kill insects in agriculture and those that carry diseases.

Decibel: A unit used to measure the loudness or intensity of a sound.

Dengue: A virus carried by the *Aedes* (primarily *Aedes aegypti*) mosquitoes causing an illness the symptoms of which are severe flu-like illness, intense bone pain, high fever, headache, rash, fatigue, and chills.

Deforestation: The cutting down of an area of forest.

Deltametrin: A chemical used to kill insects.

Deodorizer: A substance that covers up or neutralizes odors.

Dichlorvos: A chemical used to kill insects.

Dioxins: Chemical compounds produced naturally and as a by-product of industrial activities, such as, pesticide manufacturing and burning of waste—known to linger for a long time in the environment.

Distill: To separate or purify a substance by evaporation and condensation.

DNA: A molecule that contains an organism's genetic material necessary for its development, functioning, and reproduction.

Ecosystem: A system of living and non-living things interacting.

Emphysema: A disease where the lungs slowly fill up with fluid.

Endemic: A disease or organism established in an area and expected to be there for years into the future.

Endocrine System: The system of glands that produce hormones to regulate the proper functioning of the body.

Extinct: An animal or plant species that has died out completely or is no longer present in a particular area.

Famine: A severe scarcity of food in an area resulting in starvation, diseases or death.

Fermentation: A chemical reaction that converts sugars to alcohol, gases, or acid.

Fine Particles: Dust, smoke, soot and fumes that are produced from industrial processes, plowing and burning fields, road construction, and when petroleum and natural gas are burned in vehicles and power plants.

Flatulence: The presence of an excessive amount of gas in the stomach or intestines.

Flame Retardants: Chemicals added to manufactured materials, such as plastics, textiles, and furniture to make them less likely to burn and spread a fire.

Flex Fuel Vehicle: A vehicle designed to run and operate on a combination of fuels, such as gasoline mixed with alternative fuels, such as ethanol or methanol.

Formaldehyde: A strong-smelling, colorless, flammable gas or liquid primarily used to preserve human and other organic remains.

Furans: Toxic chemicals produced as a side product during the manufacture of other chemicals or found in heat-treated foods.

Gamma Radiation: Extremely high-energy invisible waves or particles capable of removing electrons from an atom.

Genetically Modified Crops: Plants grown for food, medicine, fuel or other purposes that have been artificially modified through introduction of a new characteristic, quality or condition which does not occur naturally in the species.

Glacier: A large accumulation of ice, snow, rock, and sediment that originates on land and moves down slopes under the influence of its own weight and gravity.

Glyphosate (Roundup): A chemical used to kill weeds, especially those that compete with crops.

Green Energy: Energy that comes from natural sources, such as sun, water, and wind.

Greenhouse Gas: A gaseous compound that traps and holds heat in the atmosphere.

Ground-Level Ozone: At the ground level instead of in the ozone layer, ozone becomes a secondary air pollutant.

Heavy Metals: Metallic chemical elements that have a high density and are toxic or poisonous at low concentrations.

Hepatitis: A disease that causes inflammation of the liver.

Herbicide: A chemical used to kill plants and unwanted vegetation.

Hexachlorobenzene: A chemical previously used to kill fungi and as a treatment for seeds, and is currently formed as a byproduct during the manufacture of other chemicals and pesticides.

Hybrid Electric Vehicle: A vehicle using both internal combustion and electric power sources.

Hydrocarbons: A class of chemical compounds that consist entirely of the elements hydrogen and carbon, and found mainly in crude oil.

Hydrochloric Acid: A clear, colorless, highly corrosive chemical.

Hydrochloroflurocarbons (HCFCs): Chemical compounds similar in structure to chlorofluorocarbons, including chlorine, fluorine, and carbon, but also hydrogen.

Hydrofluorocarbons (HFCs): Chemical compounds containing hydrogen, fluorine, and carbon—used as replacement for chlorofluorocarbons and hydrochlorofluorocarbons.

Hydrogen: The lightest chemical element.

Hydrogen Cyanide: A colorless and extremely poisonous liquid.

Hydrogen Sulfide: A poisonous, corrosive, flammable, explosive and colorless gas that smells like rotten eggs; often resulting from breakdown of organic matter in the absence of oxygen, such as in swamps and sewers.

Immune System: The system that protects a body against diseases.

Industrial Revolution: A period during which rural societies in Europe and America became industrialized and started manufacturing and using machines.

Intelligence Quotient (IQ): A system of assessing human intelligence.

Jaundice: Yellowing of the skin and white of the eyes often due to liver disease, such as hepatitis or liver cancer.

Lambda Cyhalothrin: A pyrethroid chemical used to kill insects.

Lead: A toxic heavy metal given off in the exhaust of vehicles that burn leaded gasoline, and in the dust of lead paints.

Lime Kiln: A furnace used for converting limestone to lime by burning.

Leukemia: A cancer of blood or bone marrow cells.

Malaria: A disease caused by *Plasmodium* parasites carried by infected *Anopheles* mosquitoes the symptoms of which are high fever, headache, nausea and chills.

Malathion: An organophosphate chemical used to kill insects.

Manganese: A silvery-gray metallic element used mainly in the manufacturing of steel, aluminum alloys and unleaded gasoline.

Mercury: A toxic heavy metal given off from burning electronic waste, garbage and biomass.

Metabolism: Includes all the chemical reactions in the body that turn food into energy and keep the body going.

Methane: A volatile organic compound found in natural gas, landfills, and garbage dumps that is about 25 times more damaging as a greenhouse gas than carbon dioxide, but doesn't stay in the atmosphere nearly as long.

Methylene chloride: A colorless, sweet-smelling liquid that evaporates easily and is used as a paint stripper and remover, to clean and degrease metals, and in the manufacture of pharmaceuticals, glues and foams.

Microcephaly: A birth defect in which a baby's head is very small and the brain is underdeveloped.

Milpa: A small field cleared from the forest to grow crops.

Mitigation: Reducing the severity of something.

Mutation: A change in the DNA of an organism.

Naphthalene: A white crystalline solid with a distinct smell that is used in fireworks and to ward off insect pests.

Nicotine: A chemical found in tobacco plants that induces addiction.

Nitric Acid: A usually colorless, highly corrosive chemical used in the manufacture of fertilizers and as a cleaning agent.

Nitrogen Oxides: Chemical compounds that consist of nitrogen and oxygen in varying combinations, produced during combustion.

Nitrogen Dioxide: A reddish-brown toxic gas with a strong odor.

Ozone: An inorganic molecule composed of three oxygen atoms that appears as a colorless or pale blue gas with a strong smell, formed naturally from lightning or ultraviolet light passing through oxygen in the atmosphere or when nitrogen oxides, volatile organic compounds, and carbon monoxide react with oxygen molecules.

Ozone Layer: A region of Earth's upper atmosphere, containing ozone, that protects living things from the sun's harmful ultraviolet radiation.

Parkinson's Disease: A degenerative disorder of the central nervous system resulting in shaking, slowness of movement, difficulty with walking, dementia, and depression.

Paraffin: A soft, waxy substance made from petroleum or coal.

Peatlands: Marshy or damp, wetland areas characterized by an accumulation of partially decayed vegetation or organic matter.

Perchloroethylene: A sweet-smelling, colorless, non-flammable liquid used to dry clean fabrics, degrease metal parts and strip paint.

Peripheral Neuropathy: A disorder caused by damage to the part of the nervous system that is outside of the brain and spinal cord, affecting sensations or movement, gland, and organ functions.

Peroxyacetyl Nitrates (PANs): Secondary air pollutants formed from volatile organic compounds and nitrogen oxides by chemical reaction in the atmosphere.

Persistent Free Radicals: Atoms, molecules or ions with unpaired electrons that make them highly reactive.

Persistent Organic Pollutants (POPs): Toxic chemicals that are transported around the world by wind and water, resistant to breakdown in the environment, and accumulate in organisms.

Pesticide: A chemical substance or mixture used for controlling, repelling, or destroying plants or animals considered to be pests.

Petroleum: A naturally-occurring clear, green, or black liquid commonly refined into various types of fuels, including gasoline, kerosene, and diesel.

Phenothrin: A pyrethroid chemical used to kill insects, such as fleas and ticks.

Photosynthesis: The process by which green plants and algae produce sugar using sunlight, carbon dioxide, and water.

Phthalates: A group of chemicals used to make plastics more flexible and harder to break.

Piperonyl Butoxide: A chemical that is added to insecticides to increase their strength.

Pneumonia: An inflammatory disease of the lungs.

Pollution: Harmful or poisonous substances that contaminate the natural environment.

Polonium-210: A rare, naturally-occurring highly radioactive element, discovered by Polish chemist and physicist Marie Curie in the late 19th century.

Polychlorinated Biphenyls (PCBs): A group of man-made chemicals used mainly as coolants and lubricants in electrical equipment and motors because they don't burn easily and are good insulators.

Polycyclic Aromatic Hydrocarbons (PAHs): Chemicals having a pleasant and distinct smell, produced when fuel is not fully burnt, such as in automobiles, incinerators, fats, tobacco, incense, and when biomass is burned in a forest.

Polymer: A large chemical made of a repeating units of smaller chemicals strung together to make long chains. Styrofoam is a polymer made of styrene molecules, so it is poly-styrene, that can be made into a foam that dries into a particular form.

Prallethrin: A pyrethroid chemical used to kill insects.

Pre-industrial: A time in human history before machines and fossil fuels were used by humans.

Propoxur: A non-pyrethroid chemical used to kill insects.

Pyrethrins: A group of chemicals compounds derived from certain types of chrysanthemum flowers and commonly used to control insects.

Pyrethroids: Man-made chemical compounds similar to pyrethrins but are more toxic and persist in the environment.

Radiation: A type of energy in the form of invisible waves or particles.

Radioactive: Giving off radiation.

Reforestation: The replanting of a forest that had been removed by cutting, fire, or disease.

Refurbish: To restore or make new again.

Reservoir: An artificial place, such as behind a dam, where a large amount of water is collected and stored for use.

Sanitary Landfill: A site where waste is disposed of by burial and kept from contaminating the environment.

Smog: A mixture of smoke, fog, and chemical fumes.

Stomata: A small opening in leaves, stems, and other parts of plants through which gases are exchanged.

Stratosphere: The second layer of Earth's atmosphere from about 8 to 30 miles high.

Styrene: A man-made chemical used extensively in the manufacture of plastics, rubber, and resins.

Sulfur Dioxide: A strong smelling, toxic gas formed as a result of burning petroleum.

Sulfuric Acid: A highly corrosive, oily, colorless liquid with many uses, including lead-acid batteries, cleaners, oil refining, and fertilizer manufacturing.

Symbiotic: A mutual benefit or dependence relationship between two or more organisms.

Tar: A solid residue of tobacco smoke.

Tetramethrin: A chemical used to kill insects.

Thyroid: One of the largest endocrine glands responsible for regulating the body's metabolism.

Toluene: A clear, water-insoluble liquid volatile organic compound used as a common solvent and as a gasoline additive.

Toxic: Capable of causing injury or death.

Troposphere: The first layer of the atmosphere that ies nearest Earth.

Trace Element: A chemical element that occurs naturally on Earth in very small amounts.

Ultraviolet Light: Short-wavelength radiation naturally present in sunlight that is not visible to the human eye.

United Nations Convention on the Law of the Sea: An international agreement that establishes the rights and responsibilities of nations in their use of the natural resources of the world's oceans and seas.

Vanpool: An agreement whereby many individuals travel together in one vehicle, usually a van, and share the cost.

Vertigo: The most common type of dizziness, where a person gets the false sensation that he/she or objects around them are moving.

Volatile: Evaporates readily at normal temperature and pressure.

Volatile Organic Compounds (VOCs): Chemical substances that evaporate easily into the surrounding air and usually have distinct scents and odors.

West Nile Virus: A viral disease carried by infected *Culex* mosquitoes causing fever, headaches, muscle pain or aches, stiff neck, rash, fatigue, inflammation of the brain and paralysis. Some birds are the primary hosts for this virus.

World Health Organization (WHO): An organization that focuses on international health issues.

Xylene: A volatile organic compound primarily used as a solvent and in the manufacture of plastics.

Yellow Fever: A virus carried by infected mosquitoes, either *Aedes* or *Haemagogus* species, that causes an illness, the symptoms of which are fever, chills, anorexia, nausea, muscle pain, and headache.

Zika: A virus carried by infected *Aedes aegypti* and *Aedes albopictus* mosquitoes that causes an illness, the symptoms of which are headaches, rash, fever, conjunctivitis (red eyes), joint pains and microcephaly, a birth defect resulting in babies with very small heads and under-developed brains.

References

Agence France-Presse. (2016, February 12). *Millions Die from Air Pollution, Mainly in China, India*. <globalpost.com/article/6732504/2016/02/12/millions-die-air-pollution-mainly-china-india.

Agency for Toxic Substances and Disease Registry (ATSDR). (1990). *Toxicological Profile for Acrylonitrile*. Atlanta GA: U.S. Department of Health and Human Services, Public Health Service.

ATSDR. (1992). *Toxicological Profile for Antimony*. Atlanta GA: U.S. Department of Health and Human Services, Public Health Service.

ATSDR. (1994). *Toxicological Profile for Acetone*. Atlanta GA: U.S. Department of Health and Human Services, Public Health Service.

ATSDR *ibid*. (1994). *Toxicological Profile for Chlorodibenzofurans (CDFs)*. Atlanta GA: U.S. Department of Health and Human Services, Public Health Service.

ATSDR. (1995). *Toxicological Profile for Diethyl Phthalate*. Atlanta GA: U.S. Department of Health and Human Services, Public Health Service.

ATSDR. (1995). *Toxicological Profile for Polycyclic Aromatic Hydrocarbons*. Atlanta GA: U.S. Department of Health and Human Services, Public Health Service.

ATSDR (1997). *Toxicological Profile for Chlorpyrifos*. Atlanta GA: U.S. Department of Health and Human Services, Public Health Service.

ATSDR (1997). *Toxicological Profile for Dichlorvos*. Atlanta GA: U.S. Department of Health and Human Services, Public Health Service.

ATSDR. (1998). *Toxicological Profile for Chlorinated Dibenzo-p-dioxins*. Atlanta GA: U.S. Department of Health and Human Services, Public Health Service.

ATSDR. (1998). *Toxicological Profile for 2-Butoxyethanol and 2-Butoxyethanol acetate*. Atlanta GA: U.S. Department of Health and Human Services, Public Health Service.

ATSDR. (1998). *Toxicological Profile for Sulfur Dioxide*. Atlanta GA: U.S. Department of Health and Human Services, Public Health Service.

ATSDR. (1998). *Toxicological Profile for Sulfur Trioxide and Sulfuric Acid*. Atlanta GA: U.S. Department of Health and Human Services, Public Health Service.

ATSDR. (1999). *Toxicological Profile for Formaldehyde*. Atlanta GA: U.S. Department of Health and Human Services, Public Health Service.

ATSDR. (1999). *Toxicological Profile for Mercury*. Atlanta GA: U.S. Department of Health and Human Services, Public Health Service.

ATSDR. (1999). *Toxicological Profile for total Petroleum Hydrocarbons (TPH)*. Atlanta GA: U.S. Department of Health and Human Services, Public Health Service.

ATSDR. (2000). *Toxicological Profile for Methylene Chloride*. Atlanta GA: U.S. Department of Health and Human Services, Public Health Service.

ATSDR. (2000). *Toxicological Profile for Polychlorinated Biphenyls*. Atlanta GA: U.S. Department of Health and Human Services, Public Health Service.

ATSDR. (2001). *Toxicological Profile for Asbestos*. Atlanta GA: U.S. Department of Health and Human Services, Public Health Service.

ATSDR. (2002). *Toxicological Profile for DDT, DDE, and DDD*. Atlanta GA: U.S. Department of Health and Human Services, Public Health Service.

ATSDR. (2003). *Toxicological Profile for Pyrethrins and Pyrethroids*. Atlanta GA: U.S. Department of Health and Human Services, Public Health Service.

ATSDR. (2003). *Toxicological Profile for Malathion*. Atlanta GA: U.S. Department of Health and Human Services, Public Health Service.

ATSDR. (2004). *Toxicological Profile for Ammonia*. Atlanta GA: U.S. Department of Health and Human Services, Public Health Service.

ATSDR. (2005). *Toxicological Profile for Naphthalene, 1-methylnaphthalene, and 2-methylnaphthalene* (Update). Atlanta GA: U.S. Department of Health and Human Services, Public Health Service.

ATSDR. (2006). *Toxicological Profile for Cyanide*. Atlanta GA: U.S. Department of Health and Human Services, Public Health Service.

ATSDR. (2007). *Toxicological Profile for Arsenic*. Atlanta GA: U.S. Department of Health and Human Services, Public Health Service.

ATSDR. (2007). *Toxicological Profile for Benzene*. Atlanta GA: U.S. Department of Health and Human Services, Public Health Service.

ATSDR. (2007). *Toxicological Profile for Lead*. Atlanta GA: U.S. Department of Health and Human Services, Public Health Service.

ATSDR. (2007). *Toxicological Profile for Xylenes*. Atlanta GA: U.S. Department of Health and Human Services, Public Health Service.

ATSDR. (2008). *Toxicological Profile for Chromium*. Atlanta GA: U.S. Department of Health and Human Services, Public Health Service.

ATSDR. (2009). *Toxicological Profile for Carbon Monoxide*. Atlanta GA: U.S. Department of Health and Human Services, Public Health Service.

ATSDR. (2010). *Toxicological Profile for Chlorine*. Atlanta GA: U.S. Department of Health and Human Services, Public Health Service.

ATSDR. (2010). *Toxicological Profile for Styrene*. Atlanta GA: U.S. Department of Health and Human Services, Public Health Service.

ATSDR. (2012). *Toxicological Profile for 1,3-Butadiene*. Atlanta GA: U.S. Department of Health and Human Services, Public Health Service.

ATSDR. (2012). *Toxicological Profile for Cadmium*. Atlanta GA: U.S. Department of Health and Human Services, Public Health Service.

ATSDR. (2012). *Toxicological Profile for Phosphate Ester Flame Retardants*. Atlanta GA: U.S. Department of Health and Human Services, Public Health Service.

ATSDR. (2012). *Toxicological Profile for Manganese*. Atlanta GA: U.S. Department of Health and Human Services, Public Health Service.

ATSDR. (2013). *Toxicological Profile for Hexachlorobenzene*. Atlanta GA: U.S. Department of Health and Human Services, Public Health Service.

ATSDR. (2014). *Toxicological Profile for Hydrogen Sulfide / Carbonyl Sulfide*. Atlanta GA: U.S. Department of Health and Human Services, Public Health Service.

ATSDR. (2014). *Toxicological Profile for Perchloroethylene*. Atlanta GA: U.S. Department of Health and Human Services, Public Health Service.

ATSDR. (2015). *Toxicological Profile for Toluene*. Atlanta GA: U.S. Department of Health and Human Services, Public Health Service.

Alegria, H.A., Bidelman, T.F., & Shaw, T.J. (2000). Organochlorine Pesticides in Ambient Air of Belize, Central America. *Environmental Science and Technology*, 34, 1953-1958.

Allen, M. (2008, July 1). *The Truth About Water-Powered Cars: Mechanics Diary*. <popularmechanics.com/cars/a3428/4271579/>

Alter, L. (2016, March 16). *If Cows Could Fly: What's in United Airlines' Biofuel?* <treehugger.com/aviation/if-cows-could-fly-whats-united-airlines-biofuel.html>

Alter, L. (2016, April 28). *Diesel is Probably Dying; Will Gasoline Powered Cars Follow?* <treehugger.com/cars/diesel-probably-dying-will-gasoline-powered-cars-follow.html>

Ambergris Today. (2015, July 10). *Belize Pushes to Go 100 percent Green with Ten Island Challenge*. <ambergristoday.com/content/stories/2015/july/10/belize-pushes-go-100-green-ten-island-challenge>.

American Geophysical Union. (2012, June 12). *Volcanic Gases Could Deplete Ozone Layer*. <sciencedaily.com/releases/2012/06/120612115920.htm>.

Associated Press. (2013, May 10). Greenhouse Gas Level Highest in Two Million Years. *NOAA Reports* (Update 2). <phys.org/news/2013-05-carbon-dioxide-atmosphere-historic-high.html>

Azar, C., Lindgren, K., Larson, E. & Möllersten, K. (2006). Carbon Capture and Storage from Fossil Fuels and Biomass – Costs and Potential Role in Stabilizing the Atmosphere. *Climatic Change*, 74, 47–79. doi: 10.1007/s10584-005-3484-7.

Badreshia, S., & Marks, J.G. (2002). Iodopropynyl Butylcarbamate. *American Journal of Contact Dermatitis* 13 (2) 77–79. doi:10.1053/ajcd.2002.30728.

Bakhiya, N., & Appel, K.E. (2010). Toxicity and Carcinogenicity of Furan in Human Diet. *Archives of Toxicology* 84 (7) 563–578. doi: 10.1007/s00204-010-0531-y.

BBC News. (2016, February 9). *Zika: Can Fish Help Stop the Spread of the Virus?* <bbc.com/news/world-latin-america-35529348>.

Beerling, D. J., & Berner, R. A. (2005). Feedbacks and the co-evolution of plants and atmospheric CO_2. *Proceedings of the National Academy of Sciences of the United States of America,* 102 (5), 1302-1305. doi: 10.1073/pnas.0408724102.

Belize and Nicaragua Logs Recovery Project. (2007). *GHG Emission Reductions Quantification Report*. Belize City, Belize: Lagacé & Legault International Inc.

Belize Ministry of Agriculture and Fisheries. (2003). *The National Food & Agriculture Policy (2002-2020)*. Belmopan, Belize.

Belize Ministry of Economic Development. (2011). *Belize Horizon 2030*. Belmopan, Belize: Barnett & Company Ltd.

Belize Ministry of Energy, Science, Technology and Public Utilities (2012). *Strategic Plan 2012-2017: Integrating Energy, Science and Technology into National Development Planning and Decision Making to Catalyze Sustainable Development*. Belmopan, Belize.

Belize Ministry of Energy, Science, Technology and Public Utilities. (2014). *Draft Final Report: Overcoming Barriers to Belize's RE and EE Potential* (Vol 1) (RG-T1886-SN2). Belmopan, Belize: Castalia Limited.

Belize Ministry of Forestry, Fisheries, and Sustainable Development. (2012). *Belize National Sustainable Development Report*. Belmopan, Belize: Institutional Development Consultants.

Belize Ministry of Health. (1996). *Diagnostic Situation on the Use of DDT and the Control and Prevention of Malaria in Belize*. Belmopan, Belize.

Belize Ministry of Health/Belize National Malaria Eradication Service. (2003). *Annual Quantity of DDT used in Belize for the Period 1985-1994*. Belmopan, Belize.

Belize Ministry of Natural Resources and the Environment/Department of the Environment. (2012). *Wealth Untold: A Glimpse of Belize's Natural Resources*. Belmopan, Belize.

Belize Ministry of Natural Resources and Agriculture/Land and Surveys Department. (2012). *Environmental Statistics for Belize - 2012*. Belmopan, Belize.

Belize Ministry of Natural Resources and the Environment/Forest Department. (2010). *IV National Report To The United Nations Convention On Biological Diversity*. Belmopan, Belize: Belize Environmental Technologies.

Belize Ministry of Natural Resources and the Environment. (2012). *Second National Communication to the Conference of the Parties of the United Nations Framework Convention on Climate Change*. Belmopan, Belize.

Belize Solid Waste Management Authority. (2011). *Waste Generation and Composition Study for the Western Corridor, Belize C.A. 2056/OC-BL – Final Report*. Belmopan, Belize: Hydroplan Ingenieur-Gesellschaft mbH.

Bell, M.L., McDermott, A., Zeger, S.L., Samet, J.M., & Dominici, F. (2004). Ozone and Short-Term Mortality in 95 US Urban Communities, 1987-2000. *The Journal of the American Medical Association*, 292 (19), 2372-2378.

Benowitz N.L. (2010). Nicotine Addiction. *The New England Journal of Medicine*, 362 (24), 2295–2303.

Benson, J. (2014, January 5). *Neonicotinoid Pesticides Not Just a Threat to Bees; Humans also at Risk*. <naturalnews.com/043399_neonicotinoid_pesticides_bees_developmental_neurotoxicity.html>.

Bernstein, J.A., Alexis, N., Bacchus, H., Bernstein, I.L., Fritz, P., Horner, E.,& Tarlo, S.M. (2008). The Health Effects of Nonindustrial Indoor Air Pollution. *Journal of Allergy and Clinical Immunology*, 121 (3), 585-591. <jacionline.org/article/S0091-6749(07)02209-9/abstract>.

Beyer, A., Mackay, D., Matthies, M., Wania, F., & Webster, E. (2000). Assessing Long-Range Transport Potential of Persistent Organic Pollutants. *Environmental Sciences & Technology*, 34 (4), 699–703. <pubs.acs.org/doi/abs/10.1021/es990207w>.

Beyond Pesticides. (2016, March 11). *Colorado Rancher to be Jailed for Pesticide Drift*. <beyondpesticides.org/dailynewsblog/2016/03/colorado-rancher-to-be-jailed-for-pesticide-drift>.

Bomgardner, M.M. (2015, February 25). *General Mills to Remove Antioxidant BHT from Its Cereals*. <scientificamerican.com/article/general-mills-to-remove-antioxidant-bht-from-its-cereals>.

Borenstein, S. (2016, April 9). *NASA: Global Warming is Now Changing How Earth Wobbles*. <yahoo.com/news/nasa-global-warming-now-changing-earth-wobbles-180545697.html>.

Borenstein, S. (2016, April 20). *Earth's Hot Streak Continues for a Record 11 Months*. <finance.yahoo.com/news/earths-hot-streak-continues-record-152700358.html>.

Botterweck, A.A.M., Verhagen, H., GoldBohm, R.A., Kleinjans, J., & van den Brandt, P.A. (2000). Intake of Butylated Hydroxyanisole and Butylated Hydroxytoluene and Stomach Cancer Risk: Results From Analyses in the Netherlands Cohort Study. *Food and Chemical Toxicology*, 38 (7), 599–605. <europepmc.org/abstract/med/10942321>.

Botzen, W.J.W., Gowdy, J.M., & van den Bergh, J.C.J.M. (2008). Cumulative CO2 emissions: shifting international responsibilities for climate debt. *Climate Policy*, 8 (6), 569-576. <tandfonline.com/doi/abs/10.3763/cpol.2008.0539>.

Bradford, A. (2015, March 4). *Deforestation: Facts, Causes & Effects*. <livescience.com/27692-deforestation.html>.

Breyer, M. (2016, March 14). *Pigeons Take to the London Skies to Tweet Pollution Data*. <treehugger.com/environmental-policy/pigeons-take-london-skies-tweet-pollution-data.html>.

Breyer, M. (2016, March 28). *Maryland to Become First State to Ban Bee-Killing Pesticides*. <treehugger.com/environmental-policy/maryland-become-1st-state-ban-bee-killing-pesticides.html>.

Caizzo, F., Ashok, A., Waitz, I.A., Yim, S.H.L., & Barrett, S.R.H. (2013). Air pollution and early deaths in the United States. Part I: Quantifying the impact of major sectors in 2005. *Atmospheric Environment*, 79, 198-208.

Caribbean Community Climate Change Center/Caribbean Carbon Neutral Tourism Programme. (2012). *Greenhouse Gas Inventory Development Report: Belize*. Belmopan, Belize: Caribbean Community Climate Change Centre.

Caribbean Community Climate Change Center. (2016, June 9). *CCCCC Awarded the Energy Globe for Biogas Laboratory Project*. <caribbeanclimateblog.com/2016/06/09/ccccc-awarded-the-energy-globe-for-biogas-laboratory-project>.

Centers for Disease Control and Prevention/National Institute for Occupational Safety and Health. (2016). *NIOSH Pocket Guide to Chemical Hazards: Carbon Black*. <cdc.gov/niosh/npg/npgd0102.html>.

Centers for Disease Control and Prevention/National Institute for Occupational Safety and Health. (2016). *NIOSH Pocket Guide to Chemical Hazards: Propoxur*. <cdc.gov/niosh/npg/npgd0531.html>.

Chan, I., Peng, S., Chang, C., Hung, J., & Hwang, J. (2012). Effects of Acidified Seawater on the Skeletal Structure of a Scleractinian Coral from Evidence Identified by SEM. *Zoological Studies* 51 (8), 1319-1331.

Cherrington, E.A., Ek, E., Cho, P., Howell, B.F., Hernandez, B.E., Anderson, E.R., & Irwin, D.E. (2010). *Forest Cover and Deforestation in Belize: 1980-2010*. <ambergriscaye.com/art/pdfs/bz_forest_cover_1980-2010.pdf>.

Cherrington, E.A, Cho, P.P., Waight, I., Santos, T.Y., Escalante, A.E., Nabet, J., &Usher, L. (2012). *Forest Cover and Deforestation in Belize, 2010-2012*. <illegal-logging.info/sites/default/files/uploads/bzforestcoverdeforestation20102012summary.pdf>.

Chow, L. (2016, February 12). *Latin American Doctors Suggest Monsanto-Linked Larvicide Cause of Microcephaly, Not Zika Virus*. <ecowatch.com/2016/02/12/larvicide-cause-not-zika>.

Chow, L. (2016, June 14). *First Mammal Goes Extinct Due to Human-Caused Climate Change*. <ecowatch.com/2016/06/14/mammal-extinction-climate-change>.

Chow, L. (2017, January 3). Scientists Say 2016 Is Hottest Year Ever Recorded. <ecowatch.com/2016-hottest-year-on-record-2176895643.html>.

Citizens Campaign for the Environment and Citizens Environmental Research Institute. (2002). *The Health Effects of Pesticides Used for Mosquito Control*. <beyondpesticides.org/assets/media/documents/mosquito/documents/citizensHealthEffectsMosqP.pdf>.

CityLab. (2016). *Not All Tree Planting Programs Are Great for the Environment: Some Tree Species Can Lead to More Ozone Production than Others, Especially iin Urban Areas* <citylab.com/weather/2014/06/not-all-tree-planting-programs-are-great for-the-environment/372849>.

Climate-KIC. (2014). *Turning on the Tap with aQysta's 'Barsha Pump'*. <climate-kic.org/case-studies/turning-on-the-tap-with-aqystas-barsha-pump>.

Cohen, A.J., Ross, A.H., Ostro, B., Pandey, K.D., Krzyzanowski, M., Künzli, N., & Smith, K. (2005). The Global Burden of Disease Due to outdoor Air Pollution. *Journal of Toxicology and Environmental Health* Part A, 68 (13–14), 1301-1307.

Colella, W.G., Jacobson, M.Z., & Golden, D.M. (2005) Switching to a U.S. Hydrogen Fuel Cell Vehicle Fleet: The Resultant Change in Emissions, Energy Use, and Greenhouse Gases. *Journal of Power Sources*, 150, 150-181. <sites.fas.harvard.edu/~scia52/Related_Material/Hydrogen_Econ/jacobson_hydrogen_cars_long2.pdf>.

Cook, J., Oreskes, N., Doran, P.T., Andregg, W.R., Verheggen, B., Waibach, E.W., &Nuccitelli, D. (2016). Consensus on Consensus: A Synthesis of Consensus Estimates on Human-caused Global Warming.*Environmental Research Letters*, 11(4), 048002.

Damstra, T. (2002). Potential Effects of Certain Persistent Organic Pollutants and Endocrine Disrupting Chemicals on the Health of Children. *Clinical Toxicology*, 40 (4), 457-465.

Dennekamp, M., Howarth, S., Dick, C., Cherrie, J., Donaldson, K., & Seaton, A. (2001). Ultrafine Particles and Nitrogen Oxides Generated by Gas and Electric Cooking. *Occupational and Environmental Medicine*, 58 (8), 511–516 <ncbi.nlm.nih.gov/pmc/articles/PMC1740176>.

Doi, H., Kikuchi, H., Murai, H., Kawano, Y., Shigeto, H., Ohyagi, Y., & Kira, J. (2006). Motor Neuron Disorder Simulating ALS Induced by Chronic Inhalation of Pyrethroid Insecticides. *Neurology*, 67 (10), 1894-1895.

Doney, S.C., Fabry, V.J., Feely, R.A., & Kleypas, J.A. (2009). Ocean Acidification: The Other CO2 Problem. *Annual Review of Marine Science*, 1, 169-192. <annualreviews.org/doi/pdf/10.1146/annurev.marine.010908.163834>.

Doyle, A. (2016, January 25). *Oslo Trash Incinerator Starts Experiment to Slow Climate Change.* <businessinsider.com/r-oslo-trash-incinerator-starts-experiment-to-slow-climate-change-2016-1>.

Edelstein, S. (2016, March 28). *Sugarcane, Modified to Produce Oil, Promises Better Biodiesel.* <greencarreports.com/news/1103102_sugarcane-modified-to-produce-oil-promises-better-biodiesel>.

Edelstein, S. (2016, March 31). *India's Ambitious Goal: All Electric Vehicles on Roads by 2030.* <greencarreports.com/news/1103162_indias-ambitious-goal-all-electric-vehicles-on-roads-by-2030>.

Edelstein, S. (2016, March 31). *Mexico Urged to Pass Strong Emissions Limits for Heavy Trucks; 20,000 deaths cited.* <greencarreports.com/news/1103173_mexico-urged-to-pass-strong-emissions-limits-for-heavy-trucks-20000-deaths-cited>.

Environment and Human Health, Inc. (2006). *The Harmful Effects of Vehicle Exhaust: A Case for Policy Change.* <ehhi.org/reports/exhaust/exhaust06.pdf>.

European Small Hydropower Association and IT Power. (2005). *Small Hydropower for Developing Countries. A Brochure Developed under the Thematic Network on Small Hydropower Project.* Brussels, Belgium.

Fargione, J., Hill, J., Tilman, D., Polasky, S., & Hawthorne, P. (2008). *Land Clearing and the Biofuel Carbon Debt. Science*, 319 (5867), 1235-1238. <science.sciencemag.org/content/319/5867/1235>.

Food and Agriculture Organization of the United Nations. (2006). *Livestock's Long Shadow: Environmental Issues and Options.* <fao.org/docrep/fao/010/a0701e/a0701e.pdf>.

Freedman, A. (2016, April 15). *March was Earth's 11th-straight Warmest Month on Record.* <mashable.com/2016/04/14/earth-11-warmest-months/#eBXXoCJFBgqK>.

Fritz, A. (2016, March 30). *Scientists Say Antarctic Melting Could Double Sea Level Rise.* <washingtonpost.com/news/capital-weather-gang/wp/2016/03/30/what-6-feet-of-sea-level-rise-looks-like-for-our-vulnerable-coastal-cities>.

Government of Belize (GoB). (2000). *Agricultural Fires Act, Chapter 204, Revised Edition 2000.* Belmopan, Belize.

GoB. (2000). *Caribbean Agricultural Research & Development Institute Act, Chapter 16, Revised Edition 2000.* Belmopan, Belize.

GoB. (2000). *Electricity Act, Chapter 221, Revised Edition 2000.* Belmopan, Belize.

GoB. (2000). *Pesticides Control Act, Chapter 216, Revised Edition 2000.* Belmopan, Belize.

GoB. (2000). *Petroleum Act, Chapter 225, Revised Edition 2000.* Belmopan, Belize.

GoB. (2000). *Public Roads Act, Chapter 232, Revised Edition 2000.* Belmopan, Belize.

GoB. (2000). *Public Utilities Commission Act, Chapter 223, Revised Edition 2000.* Belmopan, Belize.

GoB. (2000). *Sugar Cane Farmers' Association Act, Chapter 325, Revised Edition 2000.* Belmopan, Belize.

GoB. (2000). *Village Councils Act, Chapter 88, Revised Edition 2000.* Belmopan, Belize.

GoB. (2000). Water And Sewerage Act, *Chapter 222, Revised Edition 2000.* Belmopan, Belize.

GoB. (2003). *Belize City Council Act, Chapter 85, Revised Edition 2003.* Belmopan, Belize.

GoB. (2003). *Environmental Protection Act, Chapter 328, Revised Edition 2003.* Belmopan, Belize.

GoB. (2003). *Environmental Tax Act, Chapter 64:01, Revised Edition 2003.* Belmopan, Belize.

GoB. (2003). *Forest Act, Chapter 213, Revised Edition 2003.* Belmopan, Belize.

GoB. (2003). *Macal River Hydroelectric Development Act, Chapter 285:02, Revised Edition 2003.* Belmopan, Belize.

GoB. (2003). *Misuse Of Drugs Act, Chapter 10, Revised Edition 2003.* Belmopan, Belize.

GoB. (2003). *Public Health Act, Chapter 40, Revised Edition 2003.* Belmopan, Belize.

GoB. (2011). *National Energy Policy Framework: Energy Efficiency, Sustainability and Resilience for Belize in the 21st Century.* Belmopan, Belize: Tillett, Locke and Mencias.

Graves, C., Ebbesen, S., & Mogensen, M.B. (2011). Co-electrolysis of CO2 and H2O in Solid Oxide Cells: Performance and Durability. *Solid State Ionics*, 192 (1), 398–403.

GreenGeeks. (2016, February 10). *Moroccans Take the Lead in Reducing Carbon Emissions with a Big Move Toward Clean Energy.* <greengeeks.com/blog/2016/02/10/moroccans-take-the-lead-in-reducing-carbon-emissions-with-a-big-move-toward-clean-energy>.

Greenpeace International. (2007). *How the Palm Oil Industry is Cooking the Climate*. <greenpeace.org/international/en/publications/reports/cooking-the-climate-full>.

Grover, S. (2016, March 21). *World's First Double Decker Electric Bus, Now in Service*. <treehugger.com/public-transportation/worlds-first-double-decker-electric-bus-now-service.html>.

Harrison, J., Leggett, R., Lloyd, D., Phipps, A., & Scott, B. (2007). Polonium-210 as a Poison. *Journal of Radiological Protection*, 27 (1), 17–40. <iopscience.iop.org/article/10.1088/0952-4746/27/1/001/meta;jsessionid=96DEAC9572BC93C69797C256551C2DA8.ip-10-40-1-98>.

Hays, B. (2015, December 15). *Study: As Glaciers Melt and Sea Levels Rise, Earth Spins Slower*. <upi.com/Science_News/2015/12/15/Study-As-glaciers-melt-and-sea-levels-rise-Earth-spins-slower/6431450194736>.

Hecht, S.S. (1999). Tobacco Smoke Carcinogens and Lung Cancer. *Journal of the National Cancer Institute*, 91 (14), 1194-1210.

Heudorf, U., Mersch-Sundermann, V., & Angerer, J. (2007). Phthalates: Toxicology and Exposure. *International Journal of Hygiene and Environmental Health*, 210 (5), 623–634.

Iacurci, J. (2015, February 20). *Eat Less Meat Can Save the Environment, Panel Advises*. <natureworldnews.com/articles/12868/20150220/eat-less-meat-to-save-the-environment-panel-advises.html>.

Iglesias, D.J., Calatayud, A., Barreno, E., Primo-Millo, E., & Talon, M. (2006). Responses of Citrus Plants to Ozone: Leaf Biochemistry, Antioxidant Mechanisms and Lipid Peroxidation. *Plant Physiology and Biochemistry*, 44 (2-3), 125-131. <europepmc.org/abstract/MED/16644230>.

Inter-American Development Bank. (2011). *Issue Paper 07 – Existing Issues of the Environment in Belize City*. Belize City, Belize: PADECO and i.E.

Inter-American Development Bank. (2012). *BL-T018 Belize City Master Plan—Volume I Urban Development Plan. Final Report*. Belize City, Belize: PADECO and i.E.

Inter-American Development Bank/Infrastructure and Environment Sector/Energy Division. (2014). *The Energy Sector in Belize*. Technical Note No. IDB-TN-721. Washington: D.C.: Gischler, Rodriguez, Sanchez, Torres, Servetti and Olson.

International Agency for Research on Cancer. (2010, June-July). *Identification of Research Needs to Resolve the Carcinogenicity of High-priority IARC Carcinogens*. Views and Expert opinions of an IARC/NORA expert group meeting, Lyon, France. <monographs.iarc.fr/ENG/Publications/techrep42/TR42-Full.pdf>.

International Energy Agency. (2016). *Energy and Air Pollution: World Energy Outlook Special Report*. <iea.org/publications/freepublications/publication/WorldEnergyOutlookSpecialReport2016EnergyandAirPollution.pdf>.

International Energy Agency/ Photovoltaic Power Systems Programme. (2013). *Pico Solar PV Systems for Remote Homes: A New Generation of Small PV Systems for Lighting and Communication*. IEA PVPS Task 9 Report IEA-PVPS T9-12: 2012. Lysen.

Jabr, F. (2010, February 26). *Derived from Flowers, But not Benign: Pyrethroids Raise New Concerns*. <environmentalhealthnews.org/ehs/news/pyrethroids-raise-concerns>.

Japan Times. (2016, February 19). *2016 Already on Track to be the Hottest Year on Record*. <japantimes.co.jp/news/2016/02/19/world/science-health-world/2016-already-track-hottest-year-record/#.VseBoPkrLs1>.

Jerrett, M., Burnett, R.T., Pope, C.A., Ito, K., Thurston, G., Krewski, D., & Thun, M. (2009). Long-Term Ozone Exposure and Mortality. *The New England Journal of Medicine*, 360, 1085–1095. <nejm.org/doi/full/10.1056/NEJMoa0803894>.

Jones, A.P. (1999). Indoor Air Quality and Health. *Atmospheric Environment*, 33 (28), 4535-4564. <tandfonline.com/doi/pdf/10.1080/10962247.2013.801374>.

Kalyanasundaram, K., & Graetzel, M. (2010). Artificial Photosynthesis: Biomimetic Approaches to Solar Energy Conversion and Storage. *Current Opinion in Biotechnology*, 21 (3), 298-310.

Kelly, B.C., Ikonomou, M.G., Blair, J.D., Morin, A.E., & Gobas, F.A. (2007). Food Web-Specific Biomagnification of Persistent Organic Pollutants. *Science*, 317, 236-239.

Khalil, M.A.K. (1999). Non-CO2 Greenhouse Gases in the Atmosphere. *Annual Review of Energy and the Environment*, 24, 645-661. <annualreviews.org/doi/abs/10.1146/annurev.energy.24.1.645>.

Kodama, Y., Arashidani, K., Tokui, N., Kawamotoa, T., Matsuno, K., Kunugitaa, N., & Minakawa, N. (2002). Environmental NO2 Concentration and Exposure in Daily Life along Main Roads in Tokyo. *Environmental Research*, 89 (3), 236-244. <europepmc.org/abstract/MED/12176007>.

Krieger, R.I., Dinoff, T.M., & Zhang, X. (2003). Octachlorodipropyl Ether (s-2) Mosquito Coils Are Inadequately Studied for Residential Use in Asia and Illegal in the United States. *Environmental Health Perspectives*, 111 (12),1439-1442 <ncbi.nlm.nih.gov/pmc/articles/PMC1241643>.

LaMotte, S. (2016, March 11). *Stopping Zika: The GMO Mosquito that Kills His Own Offspring*. <edition.cnn.com/2016/03/07/health/zika-florida-gmo-mosquito/index.html>.

Lee, K., Xue, J., Geyh, A.S., Ozkaynak, H., Leaderer, B.P., & Weschler, C. (2002). Nitrous Acid, Nitrogen Dioxide, and Ozone Concentrations in Residential Environments. *Environmental Health Perspectives*, 110 (2), 145-150. <ncbi.nlm.nih.gov/pmc/articles/PMC1240728>.

Lévesque, B., Allaire, S., Gauvin, D., Koutrakis, P., Gingras, S., Rhainds, M., & Duchesne, J. (2001). Wood-burning Appliances and Indoor Air Quality. *Science of the Total Environment*, 281 (1-3), 47-62. <sciencedirect.com/science/article/pii/S0048969701008348>.

Liem, A.K.D., Furst, P. & Rappe, C. (2000). Exposure of Populations to Dioxins and Related Compounds. *Food Additives & Contaminants*, 17 (4), 241-259. <tandfonline.com/doi/abs/10.1080/026520300283324>.

Liu, W., Zhang, J., Hashim, J.H., Jalaludin, J., Hashim, Z., & Goldstein, B.D. (2003). Mosquito Coil Emissions and Health Implications. *Environmental Health Perspectives*, 111 (12), 1454-1460. <ncbi.nlm.nih.gov/pubmed/12948883>.

Lomnicki, S., Truong, H., Vejerano, E., & Dellinger B. (2008). Copper Oxide-based Model of Persistent Free Radical Formation on Combustion-derived Particulate Matter. *Environmental Science & Technology,* 42 (13), 4982-4988 <ncbi.nlm.nih.gov/pubmed/18678037>.

Markham, D. (2016, April 20). *San Francisco Becomes First Major U.S. City to Mandate Rooftop Solar on New Buildings*. <treehugger.com/renewable-energy/san-francisco-becomes-first-major-us-city-mandate-rooftop-solar-new-buildings.html>.

Masters, C. (2007). *Is Your Printer Making You Sick?* <content.time.com/time/health/article/0,8599,1650602,00.html>.

Meerman, J., Epting, J., Steininger, M., & Hewson, J. (2010). *Forest Cover and Change in Belize circa 1990 – 2000 - 2005*. <biological-diversity.info/Downloads/BZE_deforestation_1990-2000-2005.pdf>.

Meerman, J., McGill, J., & Cayetano, M. (2011). *Belize: National Land Use Policy for Land Resource Development*. Belmopan, Belize.

Melendez, A., Manzanero, R., & Moran, T. (2011). A Practical Guide on Agro-Ecological Methods. Benque Viejo, Belize: Friends for Conservation and Development.

Melley, B. (2016, February 18). *California Declares Massive Natural Gas Leak Sealed*. <phys.org/news/2016-02-california-declares-massive-natural-gas.html>.

Mitrovica, J.X., Hay, C.C., Morrow, E., Dumberry, M., & Stanley, S. (2015). Reconciling Past Changes in Earth's Rotation with 20th Century Global Sea-Level Rise: Resolving Munk's Enigma. *Sciences Advances*, 1 (11). <advances.sciencemag.org/content/1/11/e1500679>.

Mohney, G., & Barzilay, J. (2016, April 13). *Zika Virus Confirmed as Cause of Rare Microcephaly Birth Defect, CDC Says*. <gma.yahoo.com/zika-virus-confirmed-cause-rare-microcephaly-birth-defect-215704379--abc-news-topstories.html>.

Mooney, C. (2016, April 20). *'And then we wept': Scientists Say 93 percent of the Great Barrier Reef Now Bleached*. <washingtonpost.com/news/energy-environment/wp/2016/04/20/and-then-we-wept-scientists-say-93-percent-of-the-great-barrier-reef-now-bleached>.

Morfeld, P., & McCunney, R.J. (2007). Carbon Black and Lung Cancer: Testing a New Exposure Metric in a German Cohort. *American Journal of Industrial Medicine*, 50 (8), 565-567. doi: 10.1002/ajim.20491.

Muggli, M.E., Ebbert, J.O., Robertson, C., & Hurt, R. (2008). Waking a Sleeping Giant: The Tobacco Industry's Response to the Polonium-210 Issue. *American Journal of Public Health*, 98 (9), 1643-1650 <ncbi.nlm.nih.gov/pmc/articles/PMC2509609>.

National Aeronautics and Space Administration/Goddard Institute for Space Studies. (2016, September 12). *NASA Analysis Finds August 2016 Another Record Month*. <http://data.giss.nasa.gov/gistemp/news/20160912/>.

National Aeronautics and Space Administration/Goddard Space Flight Center. (2013, September 25). *NASA Ozone Watch*. <ozonewatch.gsfc.nasa.gov/facts/hole.html>.

National Aeronautics and Space Administration. (2016). *Climate Change: How Do We Know?* <climate.nasa.gov/evidence>.

National Oceanic and Atmospheric Administration. (2016, March). *Global Analysis—February 2016*. <ncdc.noaa.gov/sotc/global/201602>.

National Oceanic and Atmospheric Administration. (2016, June). *Global Analysis - May 2016*. <ncdc.noaa.gov/sotc/global/201605>.

New Hampshire Department of Environmental Services/Asbestos Management & Control Program Air Resources Division. (2011). *Renovation, Demolition and Asbestos: What Building Owners and Contractors in New Hampshire Should Know*. Concord, NH: New Hampshire Department of Environmental Services.

Nicholson, K. (2011). *Photocopier Hazards and a Conservation Case Study*. <cool.conservation-us.org/coolaic/sg/bpg/annual/v08/bp08-05.html>.

North Carolina State University/College of Agriculture and Life Sciences/Department of Horticultural Science. (2016). *Tree Facts*. <ncsu.edu/project/treesofstrength/treefact.htm>.

Oliveira Melo, A. S., Malinger, G., Ximenes, R., Szejnfeld, P.O., Alves Sampaio, S., Bispo de Filippis, A.M. (2016). Zika Virus Intrauterine Infection Causes Fetal Brain Abnormality and Microcephaly: Tip of the Iceberg?. *Ultrasound in Obstetrics & Gynecology*, 47 (1), 6-7 <onlinelibrary.wiley.com/doi/10.1002/uog.15831/full>.

Orr, J.C., Fabry, V.J., Aumont, O., Bopp, L., Doney, S.C., Feely, R.A., & Yool, A. (2005). Anthropogenic Ocean Acidification over the Twenty-First Century and its Impact on Calcifying Organisms. *Nature*, 437, 681-686.

Pacher, P., Beckman, J.S., & Liaudet, L. (2007). Nitric Oxide and Peroxynitrite in Health and Disease. *Physiological Reviews*, 87 (1), 315-424.

Peeples, L. (2016, January 4). *How Do You Fight The World's 'Largest Environmental Health Problem'? Harness The Sun*. <huffingtonpost.com/entry/air-pollution-cooking-stoves_us_5689fdb4e4b06fa68882b733?ir=Technology&section=us_technology&utm_hp_ref=technology>.

Pohlman, K. (2016, June 8). *Norway Becomes World's First Country to Ban Deforestation*. <ecowatch.com/2016/06/08/norway-bans-deforestation>.

Pohlman, K. (2016, June 23). *Sweden Opens World's First Electric Highway*. <ecowatch.com/2016/06/23/sweden-first-electric-road>.

Pope, C.A., Burnett, R.T., Thun, M.J., Calle, E.E., Krewski, D., Ito, K. & Thurston, G.D. (2002). Lung Cancer, Cardiopulmonary Mortality, and Long-Term Exposure to Fine Particulate Air Pollution. *The Journal of the American Medical Association*, 287 (9), 1132-1141.

Ramos, A. (2016, May 17). *1st Zika Case is North Side Belize City Woman*. <amandala.com.bz/news/1st-zika-case-north-side-belize-city-woman>.

Raupach, M.R., Marland, G., Ciais, P., Le Quere, C., Canadell, J.G., Klepper, G., & Field, C.B. (2007). Global and Regional Drivers of Accelerating CO2 Emissions. *Proceedings of the National Academy of Sciences of the United States of America*, 104 (24), 10288–93.

Renewable Energy Policy Network for the 21st Century (REN21). (2014). *Renewables 2014 Global Status Report*. <ren21.net/Portals/0/documents/Resources/GSR/2014/GSR2014_full%20report_low%20res.pdf>.

Renewables First. (2015). *What is the Difference between Micro, Mini and Small Hydro?* <renewablesfirst.co.uk/hydropower/hydropower-learning-centre/what-is-the-difference-between-micro-mini-and-small-hydro>.

Reuters. (2007, October 22). *Brazil Sugarcane Mills Agree to End Burning by '17*. <reuters.com/article/environment-brazil-cane-harvest-dc-idUSN2245758620071022>.

Reuters. (2016, March 15). *Mexico City Issues Pollution Alert over High Ozone Concentration*. <news.yahoo.com/mexico-city-issues-pollution-alert-over-high-ozone-023440484--sector.html>.

Richardson, R.B. (2009). *Human Development Issues Paper. Belize and Climate Change: The Costs of Inaction*. Belmopan, Belize: United Nations Development Programme.

RT (2016, February 9). *GMO mosquitoes Could Be Cause of Zika Outbreak, Critics Say*. <rt.com/news/330728-gmo-mosquitoes-zika-virus>.

Searchinger, T., Heimlich, R., Houghton, R.A., Dong, F., Elobeid, A., Fabiosa, J., & Yu, T. (2008). Use of U.S. Croplands for Biofuels Increases Greenhouse Gases Through Emissions from Land-Use Change. *Science*, 319 (5867), 1238-1240 <science.sciencemag.org/content/319/5867/1238.full>.

Shukman, D. (2014, November 26). *Geo-engineering: Climate Fixes 'Could Harm Billions'*. <bbc.com/news/science-environment-30197085.>

Somerville, M.F., & Liebens, J. (2011). DDTs in Soils Affected by Mosquito Fumigation in Belize. *Soil and Sediment Contamination: An International Journal*, 20 (3), 289-305. <tandfonline.com/doi/abs/10.1080/15320383.2011.560982?journalCode=bssc20>.

Soos, A. (2012, January 18). *Cruise Ship Environmental Issues*. <enn.com/top_stories/article/43872>.

Sorahan, T., & Harrington, J.M. (2007). A "Lugged" Analysis of Lung Cancer Risks in UK Carbon Black Production Workers, 1951-2004. *American Journal of Industrial Medicine,* 50 (8), 555-564 <onlinelibrary.wiley.com/doi/10.1002/ajim.20481/abstract>.

Stafford, N. (2007). Future Crops: The Other Greenhouse Effect. *Nature*, 448, 526–8. <nature.com/nature/journal/v448/n7153/full/448526a.html>.

Statistical Institute of Belize. (2013). *Belize Population and Housing Census 2010: Country Report*. <www.sib.org.bz/Portals/0/docs/publications/census/2010_Census_Report.pdf>.

Statistical Institute of Belize. (2016). *Abstract of Statistics 2013*. <www.sib.org.bz/Portals/0/docs/publications/abstract/AbstractofStatistics_2013.pdf>.

Talhout, R., Schulz, T., Florek, E., van Benthem, J., Wester, P., & Opperhuizen, A. (2011). Hazardous Compounds in Tobacco Smoke. *International Journal of Environmental Research and Public Health*, 8 (2), 613-628. <ncbi.nlm.nih.gov/pmc/articles/PMC3084482>.

Tencer, D. (2013, February 19). *Number of Cars Worldwide Surpasses 1 Billion; Can the World Handle This Many Wheels?* <huffingtonpost.ca/2011/08/23/car-population_n_934291.html>.

The Keep A Breast Foundation and Environmental Working Group. (2013). *Dirty Dozen List of Endocrine Disruptors: 12 Hormone-Altering Chemicals and How to Avoid Them*. <ewg.org/research/dirty-dozen-list-endocrine-disruptors>.

ThinkProgress. (2016, February 14). *Air Pollution Kills More People Than Malnutrition And Unsafe Sex, Scientists Say*. <thinkprogress.org/air-pollution-kills-more-people-than-malnutrition-and-unsafe-sex-scientists-say-a23aa4a689f7#.39xh5dv68>.

Thompson, A. (2014, September 2). *For Air Pollution, Trash Is a Burning Problem*. <climatecentral.org/news/where-trash-is-a-burning-problem-17973>.

Tiwari, M.K., & Dwivedi, S. (2012). Assessment of Ill Health Behaviors of Lime Kilns Workers at Maihar and Jhukehi Region of Madhya Pradesh, India. *African Journal of Environmental Science and Technology*, 6 (3), 155-159. <academicjournals.org/journal/AJEST/article-abstract/33CE80F13877>.

Treacy, M. (2015, December 9). *The Best Device for Storing Renewable Energy Could Be Made of Paper*. <treehugger.com/clean-technology/best-battery-storing-renewable-energy-could-be-made-paper.html>.

Treacy, M. (2015, April 19). *Batteries from Old Smartphones Could Light up Rural Areas*. <treehugger.com/gadgets/batteries-old-smartphones-could-light-up-rural-areas.html>.

Truth Farmer. (2013). *Killing Us Softly: Glyphosphate, Deadly Convenience*. <truthfarmer.com/2013/05/02/killing-us-softly-glyphosphate-deadly-convenience>.

United Nations Development Programme. (2012). *Draft Country Programme Document for Belize, 2013-2017*. DP/DCP/BLZ/2. <undp.org/content/dam/belize/docs/UNDP%20BZ%20Publications/Country%20Programme%20Document%202013%20to%202017.pdf>.

United Nations Environment Programme/ Division of Technology, Industry and Economics/International Environmental Technology Centre. (2010). *Waste and Climate Change: Global Trends and Strategy Framework*. <www.unep.or.jp/ietc/Publications/spc/Waste&ClimateChange/Waste&ClimateChange.pdf>.

United Nations Environment Programme. (2011). *National Environmental Summary: Belize*. Belmopan, Belize: Belize Environmental Technologies <pnuma.org/publicaciones/NES%20Final%20March%2019%202012-%20FINAL.pdf>.

United Nations Industrial Development Organization and International Center on Small Hydro Power. (2013). *World Small Hydropower Development Report 2013: Belize*. <smallhydroworld.org/fileadmin/user_upload/pdf/Americas_Central/WSHPDR_2013_Belize.pdf>.

United States Department of Labour/Occupational Safety & Health Administration. (2014). *Methylene Chloride*. <osha.gov/SLTC/methylenechloride>.

United States Environmental Protection Agency. (2015, October, 26). *Cleaning Up a Broken CFL: Recommendations for When a CFL or Other Mercury-Containing Bulb Breaks*. <epa.gov/cfl/cleaning-broken-cfl>.

United States Environmental Protection Agency. (2015, December 11). *Persistent Organic Pollutants: A Global Issue, A Global Response*. <epa.gov/international-cooperation/persistent-organic-pollutants-global-issue-global-response>.

United States Environmental Protection Agency. (2015). *Overview of Greenhouse Gases: Methane Emissions*. <www3.epa.gov/climatechange/ghgemissions/gases/ch4.html>.

United States Environmental Protection Agency. (2015). *Overview of Greenhouse Gases: Nitrous Oxide Emissions*. <www3.epa.gov/climatechange/ghgemissions/gases/n2o.html>.

United States Environmental Protection Agency. (2016). *Fix A Leak Week*. <www3.epa.gov/watersense/our_water/fix_a_leak.html>.

Velders, G.J.M., Andersen, S.O., Daniel, J.S., Fahey, D.W., & McFarland, M. (2007). The Importance of the Montreal Protocol in Protecting Climate. *Proceedings of the National Academy of Sciences of the United States of America*, 104 (12), 4814-4819. <pnas.org/content/104/12/4814.full>.

Vidal, J. (2014, November 13). *Boris Johnson Admits London's Oxford Street is One of World's Most Polluted*. <theguardian.com/environment/2014/nov/13/boris-johnson-admits-londons-oxford-street-is-one-of-worlds-most-polluted>.

Vidal, J. (2016, January 16). *Air Pollution: A Dark Cloud of Filth Poisons the World's Cities*. <theguardian.com/global-development/2016/jan/16/winter-smog-hits-worlds-cities-air-pollution-soars>.

Voelcker, J. (2016, June 14). *Nissan Takes a Different Approach to Fuel Cells: Ethanol*. <greencarreports.com/news/1104467_nissan-takes-a-different-approach-to-fuel-cells-ethanol>.

Weather.com. (2016, May 18). *April 2016 Was 12th Consecutive Warmest Month on Record, NOAA Says*. <weather.com/news/climate/news/record-warmest-april-earth-2016>.

Weaver, M. (2016, January 22). *Teacher Died from Cancer after Decades of Exposure to Asbestos*. <theguardian.com/uk-news/2016/jan/22/teacher-died-from-cancer-after-decades-of-exposure-to-asbestos>.

Westphal, M.I., & Thwaites, J. (2016). *Transformational Climate Finance: An Exploration of Low Carbon Energy. Working Paper*. Washington, DC: World Resources Institute <wri.org/publication/transformational-climate-finance>. <www.wri.org/sites/default/files/Transformational_Climate_Finance_An_Exploration_of_Low-Carbon_Energy.pdf>.

Wheeling, K. (2016, April 01). *Has Uruguay Discovered the Answer to Our Climate Change Problems?* <theweek.com/articles/615254/uruguay-discovered-answer-climate-change-problems>.

Wildtracks. (2009). *State of Belize's Protected Areas: Management Effectiveness Assessment of Belize's Protected Areas*. Belmopan, Belize.

Wigington, D. (2015). *Engineering Earth, Exposing The Global Climate Modification Assault* <geoengineeringwatch.org/e-engineering-earth>.

Worland, J. (2016). *Air Pollution Kills More Than 5 Million People Around the World Every Year*. <time.com/4219575/air-pollution-deaths>.

World Commission on Dams. (2000). *Dams and Development: A New Framework for Decision-Making*. The Report of the World Commission on Dams. London, UK: Earthscan Publications Ltd. <unep.org/dams/WCD/report/WCD_DAMS%20report.pdf>.

World Health Organization (WHO). (2009). *Global Health Risks: Mortality and burden of disease attributable to selected major risks*. <who.int/healthinfo/global_burden_disease/GlobalHealthRisks_report_full.pdf>.

WHO. (2014). *Dioxins and Their Effects on Human Health*. Factsheet No. 225. <who.int/mediacentre/factsheets/fs225/en>.

WHO/International Agency for Research on Cancer. (1986). Butylated Hydroxytoluene. *IARC Monographs on the Evaluation of Carcinogenic Risks of Chemicals to Humans*, 40, 161-206.

WHO/International Agency for Research on Cancer. (2015). *IARC Monographs Volume 112: Evaluation of Five Organophosphate Insecticides and Herbicides*. <iarc.fr/en/media-centre/iarcnews/pdf/MonographVolume112.pdf>.

WHO. (2016). *Hepatitis*. <who.int/topics/hepatitis/en>.

WHO/International Programme on Chemical Safety. (2016). *Ten Chemicals of Major Public Health Concern*. <who.int/ipcs/assessment/public_health/chemicals_phc/en>.

WHO. (2016, February 1). *WHO Director-General summarizes the outcome of the Emergency Committee regarding clusters of microcephaly and Guillain-Barré syndrome*. <who.int/mediacentre/news/statements/2016/emergency-committee-zika-microcephaly/en>.

Young, C. (2008) Belize's Ecosystems: Threats and Challenges to Conservation in Belize. *Tropical Conservation Science*, 1 (1), 18-33. <tropicalconservationscience.mongabay.com/content/v1/08-03-03-Young.htm>.

www.ingramcontent.com/pod-product-compliance
Lightning Source LLC
Chambersburg PA
CBHW040145200326
41519CB00034B/7599